WE APPRECIATE YOUR ENTHUSIASM

THE ORAL HISTORY OF Q101

WE APPRECIATE YOUR ENTHUSIASM

THE ORAL HISTORY OF Q101

James VanOsdol

Haaf-Onion
Chicago

Copyright 2012 by James VanOsdol
All rights reserved.
Printed in the United States of America.

First Haaf-Onion edition November 2012
ISBN 978-0-9839943-8-1
Published by Haaf-Onion Omnimedia
http://www.haafonion.com

This book is dedicated to the music fans and radio listeners of Chicago, whose support gave me a career and enabled me to publish this book.

CONTENTS

Foreword	xiii
Introduction	xvii
Notes about this book	xxiii
Who's Who	xxv
Friday I'm in Love: Chicago Gets an Alternative Station	1
Pass the Mic: New Blood	13
The Man Who Sold the World: The Death of Kurt Cobain	37
Twisted Christmas	41
Saturated in Guyville: The Local Music Showcase	45
Mouth for War: Rock 103.5	51
Woke up This Morning: Q101's Most Unstable Airshift	57
Not for You: Q101 vs. Pearl Jam	69
Where It's at: Jamboree and Critical Mass in the Mid-90s	75
17	85
"Here's a Ring"	89
Wendy and Bill	95
The Brew and Block Party	103
"There's Some Hoes in This House"	109
The Cow is Now	117
Mancow: In His Own Words	147
Jamboree '99	153
This Ain't a Scene: Local 101, the Payne Years	157
Anthem for the Year 2000: The New Class	163
2001: 9/11 & 94/7	171
Every Day is Exactly the Same: More Changes	181

We Are All Made of Stars: Music in the 00s	189
Evil Empire: Nipplegate	197
Woody, Tony, and Ravey	207
Electra	215
All Mixed Up: Q101 "On Shuffle"	221
The Cow is Now … Gone	231
Boulevard of Broken Dreams: The Morning Fix	237
The Lost Art of Keeping a Secret: Loose Lips and Sunken Ships	249
The Beta Brand: Q101.1	255
Sherman and Tingle	257
Icky Thumping: Q101 vs. The White Stripes	263
Portrait of an American Family: The Manno Brothers	269
St. Louis in Chicago	275
Party Hard: Drunk Shows	281
Infinite Sadness: The Last Month	287
In the End: Bonus Tracks	303
Epilogue	309
Contributors' Essays	311
Thank You	319
Acknowledgments	323

FOREWORD

BY TIM MCILRATH

Fuck You Alternative Radio. Long Live Alternative Radio.

I have a complicated relationship with commercial rock radio. Q101 largely ignored the underground-but-blossoming Chicago punk scene of the mid-90s, and, for that, Chicago punks drew a line, perhaps best summed up in the lyrics of the song "Alternative Radio" by the late Chicago punk-ska band, Slapstick:

> *"This shit's really getting to me*
> *Q101 and fucking MTV*
> *everything just seems to sound the same yeah*
> *don't care about punk rock shows*
> *great spot about the Counting Crows*
> *everything it stands for is so fucking lame*
> *Fuck you, alternative radio"*

I sang these words from the front row of more shows that I'd care to admit. But today I play in a band that benefitted from regular airplay on the same station that I was so happy to condemn as a young punk. In the fragile scene of the ecosystem in which I inhabited, underground punk and commercial radio were not unlike the feuding houses of Montague and Capulet. Like a lot of music fans, I was skeptical of the word "alternative" and all it encompassed. But in the pre-iPod / satellite radio world in the early 90s, Q101 was my link to what was happening to national music culture in the 90s. Sure I was bouncing around Cap'n'Jazz and Los Crudos shows at VFW halls and house parties, but I'd find myself at the Aragon to see Nirvana, Helmet, or Rage Against the Machine. I liked The Screaming Trees, Pantera, and those early Tool records. I loved the aggression of hardcore punk, but

I also craved the melody and big choruses of a band like Soundgarden. I was even hearing Black Flag's Henry Rollins on Q101 with his new band.

So I listened to Q101, much to the chagrin of many of my purist friends. It was right around that same time I started learning how to play guitar. Then, when some of the bands I'd seen in small punk clubs around Chicago started appearing on the airwaves, I kept listening. Bands like Rancid, Green Day, and the Offspring were suddenly embraced by alternative radio. I remained cynical about commercial radio, about music being used to sell ad space, about program directors, DJs, and shock jocks. All of it reeked of bullshit. But I couldn't ignore the music. I'd spend a little time with the college stations, but have you ever really listened to WZRD? It's mostly unlistenable. Q101 would get some stuff wrong, but they'd get stuff right too. Whether you aspired to or not, getting your song played on Q101 was a big deal, and out of reach of most local bands. If they played a song, I listened to it. If they played your song, you'd made it. The first time I heard one of our bands' songs on the radio I heard it on the same station I grew up listening to. It was special. My friends called me when it came on; my still-local family finally began to understand why I spent my formidable years in a basement with a guitar. When it became so regular that Q101 had close to ten of our songs in rotation, people were no longer surprised or congratulatory: you could expect to hear a Rise Against song at least once a day if you listened enough.

Once the band began to take off, I had the opportunity to peek behind the curtain of rock radio at stations around the country. I found that the "evil empire" was made up of mostly people just like me. Playing in a band was an extension of my adolescence, and spinning music and getting paid to talk and go to concerts was an extension of theirs. We were all beating the system in a way.

When Q101 went under years later, it was a sad, eerie silence. Like many, I was in disbelief. As a snotty teenage punk, I never thought I would go on to lament the demise of "corporate radio." I'd gone through this same cycle of emotions when Tower, Borders, and Virgin popped up and sold records out from under my favorite record stores. Fuck Tower! Shop indie! But then, one by one, they all shut their doors, and I found myself in shock, defending their existence, and missing them. Today, most physical music is sold by Wal-Mart, and even they are shrinking their shelf space every year. Sure, I had selfish reasons to be sad that Rise Against was no longer on the air in my hometown, but I wouldn't even hear a Nirvana song on the air on Chicago! These songs were still being played in other cities, but not in our town. Further, it occurred to me how ubiquitous radio still is when fans in Chicago seemed largely unaware that we had a Chicago date on our upcoming tour. Radio played a big role in how our fans found out that we were playing their town, but suddenly no station was talking about our show or other shows by bands like us.

I suspect I'm not alone when I say I have a complicated relationship with rock radio. But I owe a lot to the one in my town. Their airwaves unknowingly shaped the music fan and songwriter I would become, and then change the direction of my life. But times are hard for alternative radio. Every year, another batch of stations disappears, stations that I had a personal connection to, but none as personal as Q101.

I may have sung that Slapstick song from the top of my lungs fifteen years ago, but in the end, it would be a scorned media mogul who would deliver his own, much more potent and contractually binding version of "Fuck You, Alternative Radio."

Tim McIlrath sings, plays guitar, and writes songs for Rise Against, one of the best (and deservedly successful) bands to come out of Chicago in the past 25 years. Before Rise Against, McIlrath sang in a local band called Baxter, who put out a few underrated albums in the 1990s.

INTRODUCTION

Q101 was a mess. After years of musical identity crises, talent turnovers, management upheavals, and promotional inconsistencies, it was hard to believe anyone was still listening when Emmis Communications, the owner of the "alternative"-formatted radio station, announced the station's sale to fledgling broadcast company Merlin Media in June 2011.

Amazingly, people *were* listening. In fact, at that point, the station had started to rebound from recent years of audience abuse and erosion that included the dismissal of its highest-profile personality, a desperate format-tweaking "Hail Mary" throw by management, and an *Ishtar*-like morning show, largely regarded as one of the worst Chicago radio programs of the past twenty years.

So why were people still listening? Simply explained, the idea of Q101 was stronger than all the forces working against it. Those call letters evoked comfort, heritage, familiarity, and at least a vague understanding of what one could expect from the station. Q101 *meant something* to its audience, although it likely meant very different things to different people because the station had been maddeningly inconsistent throughout its time on the air.

In fact, the Q101 that launched in 1992 was very different from the station it transitioned into less than one year later. Q101 tiptoed its way into the alterative format after giving up on a safe adult contemporary format, which had earned it ample market familiarity and relative financial success. Upon its launch, "Chicago's New Rock Alternative" (as it was called back then) had heritage rocker WXRT, an "Adult Album Alternative" station, in its sights. "XRT" had long been established in the market as the coolest commercial station in town, anchored by disc jockeys who knew the music they played and actually picked the songs they played on the air (the rarity of such a thing cannot be overstated).

The play for XRT's audience was misguided and confusingly executed. How else can one explain post-format-change playlists that mashed up adult contemporary ("AC") holdovers like Phil Collins, Don Henley, and John

Mellencamp and poppier, '80s "alternative" bands like R.E.M., The Cure, and Depeche Mode?

As Q101 worked to find its footing, the 'quote-unquote' alternative scene was exploding in the aftermath of major album releases, such as Nirvana's *Nevermind* (1991), Pearl Jam's *Ten* (1991), and Jane's Addiction's *Ritual De Lo Habitual* (1990). After a few too many Sting records, Q101 brass came to realize that there was an audience hungry for a station that would support, curate, and take ownership of alternative music and its attendant cultural groundswell. Henley had to go, and Cobain needed to be welcomed.

KROQ in Los Angeles, the most respected "modern rock" station in North America, had planted its flag on Mount Alternative long before Q101 laced up its hiking shoes. For well over a decade prior to the release of *Nevermind*, KROQ had been routinely giving airtime to "underground" and fringe artists. Once alternative music started to find its way in front of a mainstream audience, alternative stations in search of touchstone guidance, including Q101, emulated KROQ's esthetic vibe, and playlist. It seems absurd that a station airing a format spun out of an underground music culture, headquartered in the third largest market in the U.S., gave more credence to decisions made at a radio station two thousand miles away, than it did to trusting its own instincts. However, while the idea was absurd, it worked.

In radio, it doesn't matter where the ideas come from, as long as they pay off. Once Q101 chucked all remaining traces of adult contemporary and actively started embracing the alternative format's marquee artists—Nirvana, Pearl Jam, Red Hot Chili Peppers, Nine Inch Nails, and Smashing Pumpkins—the station looped into a solid groove. Ratings went through the roof, and the station frequently exceeded its budget goals. Q101 landed on the right format at the right time.

Concurrent with Q101's growth was an increased amount of national attention given to the Chicago music scene. After the commercial success of Smashing Pumpkins' *Siamese Dream* (1993), the critical buzz behind Urge Overkill's *Saturation* (1993), and Liz Phair's *Exile in Guyville* (1993), many thought of the Chicago scene with the same sense of reverence previously afforded to Seattle, which had given the world Nirvana, Pearl Jam, Soundgarden, and Alice in Chains. Suddenly, fat-cat record executives looking for Windy City artists to call their own were courting every band around Chicago— both good and bad.

Q101 benefitted from peaking at the same time as the Chicago music scene's national moment, even though (embarrassingly) KROQ first aired local-to-Chicago, mid-90s bands Veruca Salt and Smoking Popes, who were the "next big things."

As alternative music was Q101's understood identity, the disc jockeys were its public face. Not long after the format flip to alternative, Q101 brought on experienced DJs like Steve Fisher and Robert Chase to anchor key "day-parts" (*shifts*, in radio lingo). Chase and Fisher—and later on, personalities like Brooke Hunter, Zoltar, Brian "The Whipping Boy," and Samantha James—contributed to Q101's "stationality" in the 1990s in much the same way as the music of Nine Inch Nails and Pearl Jam defined it.

When the alternative movement started to wane in the mid-to-late '90s, as all music movements do, Q101 became vulnerable. Not helping matters was a hungry, edgy, *dangerous*-sounding station across the metaphorical street that was looking for ways to destroy Q101. Rock 103.5 (WRCX) was an "active rock" station built on attitude, booze, and relentless guerrilla marketing. For a couple of years, the two stations were embroiled in a death match, with no guaranteed winner.

Rock 103.5's core strength was Mancow's Morning Madhouse, an in-your-face, if-you-don't-listen-you're-gonna-miss-something, ratings juggernaut of a morning show, hosted by Erich "Mancow" Muller, one of the 1990s' true radio rock stars. Q101 continually tried, and repeatedly failed, to stall Muller's growth in morning drive. The station launched five major morning shows between its sign-on and 1998; none of them could touch Mancow's ratings or street buzz.

When Mancow's show ended every day, Q101's anorexic morning ratings would rebound, as soon as Q101 and Rock 103.5 went head to head with their music formats. The mass appeal of the alterative format, at the time, was more suited to office listening, which helped drive the Arbitron numbers up. It was as if the two stations traded audience shares every day at ten a.m. (or later, depending on when Mancow actually walked out of the Rock 103.5 studio for the day).

The Mancow problem was real. Q101's inability to gain traction in morning drive was real. Q101's solution? Embrace the enemy. In 1998, Mancow Muller announced that he was leaving Rock 103.5 to become Q101's new morning host—for a three-year deal that included a salary of reportedly near $3 million a year. Mancow's arrival at Q101 was a legitimate game changer: it forced WRCX to change its format, it brought immediate ratings increases to Q101's morning drive and overall station numbers, and it altered the station's internal culture for good.

For those who worked at Q101 prior to Mancow's arrival, the cultural difference was obvious and immediate. Suddenly, there were strippers and porn stars hanging outside the studio. There was drinking in the hallways. There were words, phrases, and jokes yelled down the hallways that would have made a hardened HR administrator admit defeat and change careers.

The other side of the culture shift was the higher profile Mancow brought to the station. The media had a love/hate affair with him and his show, but they couldn't stop reporting on what he was doing and saying. The level of stars—actors, musicians, and athletes—coming by the Q101 studio radically increased, as well. Mancow was a legit star on the morning radio landscape, and because of that, he amplified everything around him.

The tone of Mancow's show skewed way more "dude" than the mostly female-leaning Q101 had ever considered. With that change, the station tweaked its music accordingly. Gone were the days of Tori Amos and Blink 182, as the next wave of rock bands, including mook-rockers Creed and Limp Bizkit, wrested control of the playlist.

Following Mancow through Q101's front door were former Rock 103.5 personalities, including Sludge and Chris Payne. Both might've had an awkward first week or two on the air, but they quickly became defining personalities for the station. Payne, for example, earned the distinction of being the last voice heard on the air when the station signed off in 2011.

In the 1990s, morning show turnover became a punch line and a sore spot for station management and employees. In the '00s, with Mancow firmly in place, morning show turnover was no longer an issue. Instead, management turnover became one of the station's most significant challenges. The job of Program Director changed hands seven times between 2001 and 2011. With each Program Director change, came staff shakeups, disc jockey firings, and format tweaks that, on the surface, appeared driven more by program director vanity than by an actual understanding of the audience.

With the managerial tweaks and changes, came inconsistencies in how management regarded Mancow. Most Program Directors treated Mancow's relationship with the station as "church and state;" the two were to exist separately. That segregation led to position statements that sounded like apologies: "Mancow mornings, Q101 music all day."

Air-staff turnover became a lot more common in the 00s, as different Program Directors came and went. On a positive note, the turnover and changes yielded some of Q101's most remembered disc jockeys: Electra. Ryan and Kevin Manno. Fook. Alex Quigley. Sherman.

Q101 made one of its most radical changes when it went "on shuffle" in the mid-00s. The "soft format change" followed the trend set by "JACK" stations that were popping up all over the country. The hallmark of the scattershot JACK format was its positioning statement, "We play anything." In truth, the JACK format played a lot more songs from a lot more genres than other formats did, but those songs were hardly adventurous or groundbreaking when held up against any music fan's iPod. A "Hail Mary" throw by all accounts, "Shuffle" allowed Q101 to widen its playlist to encompass a much broader definition of alternative music, opening the airwaves up to music from the 70s through to the present day. The shock of hearing Bob Marley running into Disturbed, then segueing into Dexy's Midnight Runners was fun when it first started, but it had become scarcely listenable when it was inevitably pulled off the air.

As the music was shuffling, so too was the morning drive slot. After eight years with Q101, Mancow's contract wasn't renewed in 2006. The station had a plan to replace him; it just happened to be a really bad plan.

The Morning Fix, a scripted, comedy, morning show ensemble, was a disaster. The approach, a *Daily Show* for alternative-music-listening, young professionals, failed on all levels; most significantly, it was a comedy show that was rarely funny. Despite an honest and laudable effort from anchor Alan Cox, the show was uncomfortable to listen to, and a total drag on the ratings—especially compared to Mancow's much-stronger and historically significant run.

As the years progressed, Q101 became the victim of an industry freefall. Revenues were down and more personal and accessible mediums, from satellite to mobile phones, and from Pandora to podcasts, seduced listeners.

Rumors of a sale hounded the station for years, leaving many on payroll to wonder when that shoe might eventually drop. In June 2011, the announcement came that Merlin Media had bought Q101 from Emmis (along with two other stations owned by the same company, WLUP in Chicago, and WRXP in New York City).

If it learned nothing else in its 19 years on the air, Q101 had retained one of the most important lessons in rock and roll: it's better to burn out than fade away. During its last week on the air in July 2011, the station went out with a bang, rather than a whimper.

During its final week, the programming was an everything-goes, all-bets-are-off affair from 6 a.m. until 10 p.m. (sadly, the station was automated between 10 p.m. and 6 a.m.). The full-time air personalities (Sherman and Tingle, Electra, Tim Virgin, and Pogo) played and said whatever they wanted. What listeners heard was real. It was honest. It was engaging. With no mandates coming from a corner office, the station sounded like the broadcasting ideal—disc jockeys curating music they loved and not restricting their talk breaks to upholding the whims of a ratings-gaming program director. It was refreshing, but it could only exist in the context of a "nothing matters" commercial radio environment.

For its last day on the air (July 14, 2011), Q101 turned the broadcast day into a homecoming. Old employees and disc jockeys stopped by or called in to the studio. They recounted memories -- good, bad, funny, and awkward. I was out of town that day, but I was invited to call in twice. My first time on the air, I talked on the phone with Tim Virgin and Brian Sherman. We laughed as I shared my most humiliating moment—finding out from a listener that Q101 was going to fire me (more on that later).

I was on the air for a second time that day, in the last hour of the station's life. I felt honored to appear on the air with Chris Payne, the long-running host of Q101's Chicago music show, *Local 101*. I hosted the show for five years, from 1995-2000, and I had always considered Payne a kindred spirit. As Chris and I talked, I felt emotional about Q101's exit even though, intellectually, I knew that my life would go on happily without ever hearing another Beastie Boys song on the radio. It was a strange contradiction: I hated the idea of losing Q101, despite the fact that I'd found it maddening to listen to, more often than not.

And maybe that feeling—that hard-to-explain, intangible, idea of Q101—is what kept people listening through Q101's most questionable programming and decision-making. Q101's presence at 101.1 on the FM band offered comfort and security to us. You could depend on it being there, and on feeling some sort of connection to it, regardless of who or what was on the air. Maybe you first made out to a song Q101 played. Maybe the station sponsored a concert that changed your life. Perhaps a DJ said something that stayed with you over the years. Whatever the case, it was an easy "go-to" on the dial, always offering the promise of being better than it was.

NOTES ABOUT THIS BOOK

We Appreciate Your Enthusiasm is an "oral history," meaning that the narrative is largely built from the stories, thoughts, and recollections of the people I interviewed. It tells a story, and in many cases, it will seem as though some interviews occurred with multiple people at the same time, although I conducted all interviews individually.

Memories blur as the years progress. The stories shared for this book were told by the interviewees to the best of their recollections, and I did my best to fact-check them. If you notice inconsistencies or gaffes, please forgive the oversights.

Some people were simply not interested in contributing. Others said they would talk, and then stopped returning my messages. Some colleagues wouldn't take my calls at all. With that, I feel honor-bound to say that I made every attempt to interview all the key players from the station for this book.

On that note, there's the issue of Mancow. From 1998-2006, Mancow's Morning Madhouse was a massive presence on Q101 (to say nothing of it also being a nationally syndicated program). Mancow Muller's contributions to Q101 were many, and I couldn't imagine *not* including his thoughts in this book.

I knew there would be challenges in getting him to agree to an interview. When news of this book first broke in 2011, Mancow told Chicago media journalist Robert Feder, "I don't live my life looking in the rearview mirror. I like James VanOsdol a lot. I hear he's writing a book. I was the biggest player in the story, and I wouldn't read a book about that. I don't understand why anyone cares."[1]

After some back and forth, Mancow generously agreed to talk. He did so in an attempt to "set the record straight." The results appear in Chapter 16 (no skipping ahead!).

It was difficult to know how to include my own experiences at Q101 in the pages that follow. I worked at the station on three separate occasions: 1993-2000, 2006-2007, and 2009-2011. I saw, did, and experienced a lot in my time there.

---

[1] Robert Feder, "Mancow Muller on Q101's Demise," *Time Out Chicago*, August 17, 2011, 17.

However, I'm also the author of this book. It didn't feel right to insert myself into the story; so, I didn't—for the most part.

One obvious exception can be found in Chapter Five, which is specifically about *The Local Music Showcase (Local 101)*, a show I started hosting in 1995.

Then, there was the time Q101 fired me in 2007. Because of the insanity involved in that situation (just you wait—you'll be aghast), I took a more subjective approach to the story's set-up in Chapter 29.

A handful of quotes in this book reference me by name. In each instance, the interviewee's quote was included because it was important to the narrative, not because I was looking for any additional glory, credit, or attention.

Finally, given the volume of names being tossed about in the book, I've tried to make reading the book easier by identifying the interviewees by their most significant or memorable Q101 job titles when their names first appear in each chapter.

WHO'S WHO

Dave Ball: Air personality

Susan Banach: Air personality

Jaime Black: Producer (*Local 101, Sonic Boom, TBA*)

Mike Bratton: Creative Director; Production Director

Robert Chase: Air personality

Nikki Chuminatto: Air personality

AJ Cox Rudolph: Sales Promotion Manager

Alan Cox: Air personality

Dana Lucas: Air personality

Jeff Delgado: Promotion Assistant

Natalie DiPietro : "Director of Fun & Games"

Chuck Ducoty: Vice President/General Manager

Mike Englebrecht: Promotions Assistant; National Sales Coordinator; Integrated Marketing Manager

Spike Eskin: Assistant Program Director/Music Director

Steve Fisher: Air personality

Fook: Air personality

Freak: Air personality

Bill Gamble: Program Director/Regional Vice President of Programming

Mark Goodman: Air personality

Phil "Twitch" Grosch: Air personality; Webmaster

Kyle Guderian: Operations Manager

Brian "Sludge" Haddad: Air personality

Chuck Hillier: General Manager

Brooke Hunter: Air personality

Jacent Jackson: Assistant Program Director/Music Director

Tim Johnson: Marketing Director

Jeff "J Love" Karlov: Producer; Air personality; Interactive Account Manager

Abe Kanan: Creative Producer

Tom Kokinakos: Research; Producer

Vance Koretos: Research

Tisa Lasorte: Director of New Media; Brand Manager

Carla Leonardo: Air personality

Bill Leff: Air personality

Dan "Bass" Levy: Air personality; producer

Jen "Jameson" Longawa: Air personality

Alex Luke: Program Director

Jim "Jesus" Lynam: Executive Producer (*Mancow's Morning Madhouse*); Air personality

Madison: Air personality

Kevin Manno: Air personality

Ryan Manno: Air personality

Rey Mena: Local Sales Manager; Marketing Director

Erich "Mancow" Muller: Host, *Mancow's Morning Madhouse*

Robert Murphy: Air personality

Marv Nyren: General Manager

Eric "Shark" Olson: Air personality

Brian "The Whipping Boy" Paruch: Air personality

Christine "Electra" Pawlak: Air personality

Chris "Payne" Miller: Air personality

Brian Peck: Air personality

Christian "Cap" Pedersen: Promotion Assistant; National Sales Coordinator

Pogo: Air personality

Alex Quigley: Air personality

Ravey: Air personality

Jon Reens: Research; Promotions Assistant

Dave Richards: Program Director

Tim Richards: Program Director

Midge "DJ Luvcheez" Ripoli: Technical Producer (*Mancow's Morning Madhouse*)

Ken "Al Roker, Jr." Smith: Air personality; Sports Director; Producer

Michelle Rutkowski: Promotions Assistant; Air personality

Keith Sgariglia: Promotions Manager; Interactive Manager; Air personality

Brian "Sherman" Sherman: Air personality

Katie (Jaski) Sherman: Promotions Assistant; Senior Project Manager

Mary Shuminas: Music Director/Assistant Program Director

Alan Simkowski: Director of Market Development

Bobby Skafish: Air personality

Wendy Snyder: Air Personality

Ned Spindle: Creative Director

Mike Stern: Program Director

Stoley: Air personality

Joey "Just Joey" Swanson: Air personality

Lance Tawzer: Air personality

Steve Tingle: Air personality

Samantha "Samantha James" Tuck: Air personality

Tim Virgin: Air personality

Woody: Air personality

Zoltar: Air personality

FRIDAY I'M IN LOVE: CHICAGO GETS AN ALTERNATIVE STATION

CHAPTER ONE

When Q101 signed on as an "alternative" station in 1992, the station walked away from its established history in the Chicago radio market as an "Adult Contemporary"-formatted FM outlet (Adult Contemporary is the format known for dentist-office friendly songs like Elton John's "I Guess That's Why They Call it The Blues" and "Make Me Smile" by Chicago).

Leading up to that point, from the 1980s' fading months through the beginning of the 1990s, Q101's conservative, mid-40s demographic-leaning mix of music was good enough to keep the station afloat in the ratings.

It didn't hurt that the station had one of Chicago's most well known air personalities at the time anchoring the station in morning drive: Robert Murphy, host of the "Murphy in the Morning" show.

ROBERT MURPHY, *Air personality*

I arrived there in January of 1983. I came in; I'd been working in Charlotte, North Carolina. They had offered me the job previous, and I was unable to get out of a contract … when I finally was able to finagle my way out of the contract, I was happy to come there. I knew I needed to come to Chicago; I knew it was where I needed to be. I didn't get exactly what I wanted to come there, but I knew it was time, or I was going to be stuck in Charlotte for a lot longer.

When I first got there, they were still playing Peter, Paul, and Mary's "Leaving on a Jet Plane" and Barry Manilow, but it was right at this time that the music like Pet Shop Boys, Tears for Fears, and Duran Duran, and all the British groups came in, and I think we were kind of a leader in that. That's what we were playing; it was all kind of strange and new at that time. I think that kind of music defined the next ten years that I spent there.

BILL GAMBLE, *Program Director*

I got to Q101 in 1988, and at the time, NBC owned it. Emmis (Communications) bought us the end of '88.

ROBERT MURPHY

Emmis was a radio corporation; GE had bought up NBC; they were a light bulb corporation. It was the start of what has happened today, those big conglomerates and all that. It was nice to work for a network O&O (owned and operated) at that time. I didn't know who Emmis was at that time, but they were radio people; they came in and they were expanding, and that was a good move for them.

CHUCK HILLIER, *General Manager*

Emmis entered in 1988. They had bought all of NBC Radio, which was pretty cool. NBC had already sold WMAQ six, seven, or eight months beforehand. (Owner) Jeff Smulyan actually bought NBC radio; this is pretty remarkable, in retrospect. NBC had never made a nickel on the FM dial in the market of Chicago. They had never made a profit, and there were some people with the old Q101 under NBC ownership. Revenues were at a high of six million dollars, and none of it went to the bottom line.

BILL GAMBLE

We were in an Adult Contemporary battle with "Clear" at the time, which is now the Mix, and FIRE, which is KISS now. Basically, we were all forms of Adult Contemporary, just basic AC radio stations. All three radio stations were following the same strategy. We all had morning shows. We all spent literally a couple million dollars a year on television, and we were all sitting with about a two share (a share is a station's average quarter hour audience represented as a percentage of all available listeners during a specific time frame). We all kind of thought there had to be a better way to do this. We knew we had to do something different; we didn't know what we needed to do.

CHUCK HILLIER

There were three stations, those two stations, plus Q101. All were playing identical music. If you were listening after 10 o'clock in the morning, you couldn't tell which of the three stations you were listening to. The distinguishing characteristic, of course, was "Murphy in the Morning." It was what we were most known for, and it helped give us a hipper edge … perceptually, people viewed us as a little bit cooler. Any time that multiple stations play the same music, it's an opportunity to go right to the other guy when people hear a song they don't like, and there were two other guys. Not only were there three stations carving up two stations' worth of total audience, but the revenues were slim.

MARY SHUMINAS, *Music Director*

At the time, Q101 was AC, and we had a morning show called Murphy in the Morning. It was probably the equivalent of (WTMX morning show) Eric and Kathy, as far as appeal. The morning show was really huge in the 80s, but it was past its peak. The rest of the station was lower than the morning show, playing stuff like Don Henley and Rod Stewart …

ROBERT MURPHY

(Ratings were) down, from what it had been. There were other stations starting to make some noise. The Mix was just getting started back then; they were (W)CLR, licensed to Skokie. Now they were next door. I could see them through a window now. You still had WLIT. I don't remember who was on there, to be honest; they were just playing AC music.

CHUCK HILLIER

If there's one underlying belief that I have as a salesperson, or as a Vice President and General Manager, or any of the other opportunities I've been lucky enough to be involved with in broadcasting, I believe wholeheartedly in that recent (Jack) Trout quote, "differentiate or die." If you're identical to somebody else, people have the opportunity to consider you irrelevant. Q101's differentiation was cool promotions, Murphy in the Morning, and that was about it.

MARY SHUMINAS

If I remember correctly, I think we were talking about a format change at the beginning of '92

CARLA LEONARDO, *Air personality*

About six months before we changed formats, I went into Bill because Q101 was slipping in the ratings terribly. We weren't getting a good piece of the pie; nothing was happening. There was competition from everywhere. To me, it just wasn't making any sense.

BILL GAMBLE

Everybody had an opinion. I can't really tell you that there was anybody waving the alternative flag at that time. So, we did a research project; Joint Communications, John Parikhal and his group, did it. John, as I remember it, didn't see an alternative point of view but saw that some modern music could be mixed in with what Q101 was playing at the time, which might become a format.

He said, "Jeez, what would this format consist of?"

(I said), Well, maybe you play the Cure, but you also play Phil Collins. What we're doing right now isn't working, so we're willing to try anything.

MARY SHUMINAS

We had done a research project. I remember having conversations with Bill in '91 and '92. The first Lollapalooza was absolutely huge, and no one was playing that music. XRT was, but you would hear a hot song like Jane's Addiction, like, once a day, or four times a week

CHUCK HILLIER

Very few people know, or may recall, that there were actually two options for us. The format finder that (consultant) John Parikhal did for us unveiled two opportunities. One was an adult urban radio station. How about that? That could've been Q101.

The other was alternative, adult alternative. Not modern rock, but adult alternative, as we had researched it

MARY SHUMINAS

I don't even remember the other genres of music we researched, but there was one, and it was probably some really conservative alternative, like U2, R.E.M., and the Cure. And it just came back through the roof. There was a decision that we were going to go in that direction.

ROBERT MURPHY

I knew something could happen; I had no idea what it might be, although I was certainly privy to the information of what was going on. It wasn't like they caught me by surprise.

BILL GAMBLE

We had a fair amount of revenue on the station and we were a female-based radio station. The goal at the time was to keep our female base and to keep the revenue we had. We also had a contract left for a while with Murphy, and no one really had an appetite to say, "Hey, Robert, here's a big check to go away," because we were still making money from him, selling the show.

VANCE KORETOS, *Research*

I do remember in June of 1992, Mary sending me to the music store to buy CDs from artists like the Cure, Depeche Mode, and Red Hot Chili Peppers. Plus, I noticed a lot of stacks of CDs in her office of artists we didn't play at the time. So, I had a hunch that something was going to change.

MARY SHUMINAS

I remember Corporate saying, "Oh, it's not going to sound anything like (Los Angeles Alternative station) KROQ, is it? You're not going to be playing 'Detachable Penis' (King Missile)?" We're, like, "Oh, no, no, it's not going to be anything like that." Of course, in my mind, I'm thinking, "Yes, I hope it's just like that."

CHUCK HILLIER

I don't think there was anybody that was recalcitrant on the part of Corporate. I loved how those guys loved radio. To boot, they were confident, legitimate, businesspeople. They wanted to provide a product that would reach out to the greatest possible share of the marketplace

CARLA LEONARDO

(Bill) calls me into his office and under penalty of death says, "We're switching formats." He wanted me to help him get the music together. I had to get pronunciation guides together for our jocks who were Adult Contemporary, to let them know that it was the La's, not the L.A.'s.

MARY SHUMINAS

Carla, gosh, she was really key to the launch of the radio station. She and I worked together as far as exposing Bill to the alternative music. We would put together cassette tapes for Bill to listen to in his car on the way to work, and of songs for Bill to say "yes" or "no" to as far as our gold library, before we signed on. She also put together stuff for the jocks as far as educating them on the bands because we had kept a lot of the staff from the Adult Contemporary format. She put together a little encyclopedia catalog of artists that we were playing. She was very passionate about music and very knowledgeable about the music format.

ROBERT MURPHY

Bill Gamble took me over to the saloon and showed me the playlist, and there was the Cure. I was happy about it because I really was thinking that the station had become unpardonably unhip and was looking forward to this. I think, at the time, Bill Gamble said, "You realize this could be the end of you." Not like, this is a threat, or anything; just this may change things. And I said, "Yes, I do know that," and I did.

CHUCK HILLIER

Here's how different the times were. We shared what we were going to do the following Friday, on a Monday, in a closed-door meeting. We shared with at least the on air staff, and maybe the marketing staff, and maybe a couple or two sales people, and certainly the department heads. We told everybody to respect that this had to be confidential.

I look back over a decade later ... I believe that everybody kept their word and just shut their mouths until the unveiling. That could never happen today.

BROOKE HUNTER, *Air personality*

We had this big station meeting, and we were flipping to alternative. We were all so excited. I still remember the very first song ever played, "Friday I'm in Love" by the Cure. Carla Leonardo was the one who kicked off the format. She was totally, totally, nervous.

CARLA LEONARDO

Unbelievably nervous. It was crazy. It was a massive switch. There had been lots of talk about it; there had been lots of anticipation. All the big 'muckety mucks' were there, all the big people from Emmis.

REY MENA, *Local Sales Manager; Marketing Director*

I happened to be at the meeting with all the sales people and programming when they launched it in front of all the buyers, at this luncheon that they had. It was pre-cell phone, so I was in the hallway, right at the doorway, looking at them on the podium, and then they gave me the signal and there was somebody in the hallway on a pay phone. I gave them the signal. They were connected to the studio, waiting for the signal so that they could flip the station and play "Friday I'm In Love."

CARLA LEONARDO

It was probably the highest moment in my career, switching over with the Cure's "Friday I'm in Love."

BROOKE HUNTER

It was very exciting. When Q101 was doing its Adult Contemporary format, none of us was a big fan of it, because we were all 20-something kids working there in Programming and Research.

VANCE KORETOS

When Q101 did change format and had the new slogan of "the New Rock Alternative," we had an odd mix of core artists we were playing when the format started. Artists like Eric Clapton, Steve Winwood, and John Mellencamp mixed with The Cure, Depeche Mode, The Breeders, and Red Hot Chili Peppers.

REY MENA

They were straddling the fence; they wanted to be alternative and went back and forth on how to name it. It was Chicago's New Rock Alternative, because they wanted to straddle the New Rock and the word alternative, but then they were playing everything from Don Henley to the Cure.

ROBERT MURPHY

I don't know if most people remember, Q101 did not go from AC to Alternative; they flipped from AC to a Triple-A format, much like at WXRT. In the beginning, we were still playing old songs by Bob Seger and deep cuts by Steely Dan, along with the Cure and newer music.

CHUCK HILLIER

Chicago's New Rock Alternative—that term—you cannot believe the excruciating research that took place that ultimately led to those three words.

MARY SHUMINAS

It was so conservative because of what Corporate wanted. We were working with a consulting company. I think the hardest song on the playlist at the time was Gin Blossoms' "Hey Jealousy." At the time, we were still playing John Mellencamp, Don Henley ...

CARLA LEONARDO

I remember Chuck Hillier coming in saying, "Are you sure Chicago isn't an alternative band we can play?"

ROBERT MURPHY

I actually enjoyed the music—everybody thought I must've hated the music—no, I was cool with it, everything was fine. I was 42 years old. The new staff was all 19-22, they weren't going to give me a chance to be a purveyor of that music, and I understood that also.

BILL GAMBLE

When I look back now, it was probably more of a Modern AC. We all like to point to the date; it was "Friday I'm in Love" by the Cure ... it was in 1992. But that radio station evolved pretty quickly. I think the interesting thing is that we really did a pretty good job of listening to the audience. They were pretty quick to say, "It's pretty cool that you're playing Depeche Mode, and REM, and the Cure, but if you play Don Henley, I'm going to put a stake through your eye." We listened to the audience and the music mix kept getting refined and refined, and ultimately, it led to being Q101. It was a rapid evolution.

MARY SHUMINAS

A lot of the artists we were playing were XRT artists like XTC and Morrissey. It was also with mainstream AC artists. Eventually, that stuff went away. We installed cassette recorders, we had listener lines, and we actually transcribed everything for the first couple of months. The listeners were coming back saying, "Tom Petty doesn't fit," or "How could you call yourself an alternative station and not play Pearl Jam or Nirvana?"

REY MENA

It was an AC station trying to pose as a cool alternative rock station, not knowing what alternative rock was. You had this format that made absolutely no sense. They weren't playing Nirvana because it was too edgy. They weren't playing Pearl Jam, because it was too edgy, but they were playing the Cure; they were playing a bunch of 80s-type music.

BILL GAMBLE

Five months after we signed on, we were a lot different. We just realized that that mix wasn't right.

MARY SHUMINAS

A quintessential song on Q101 was "Under the Bridge" (Red Hot Chili Peppers); that's what the station sounded like initially, mixed in with this other stuff.

ERIC "SHARK" OLSON, *Air personality*

I do remember being a huge fan of 99.1 WHFS (Washington, D.C.) when I was working back east, and on my summer 1992 road trip out west, I drove thru Chicago on my way to see the parents in Madison … the very first song I heard on Q was "Girlfriend" by Matthew Sweet. Even though I was still working on CHR (Contemporary Hit Radio) stations, I was so fucking excited that Chicago was getting an alternative station and I thought someday, maybe, just maybe, I'd be on a station like this.

BILL GAMBLE

I think the radio station and the company and everybody did a pretty good job of looking, and seeing what was happening. One of the things that surprises people when they talk about Q101, at least the time that I was there, is yeah we did research. We researched our songs. Songs that we were playing, we would research. We always looked at women. We found that finding those big alt-rock records that women loved made us so big. Because guys are easy, "Oh, okay, it rocks." Fine. We didn't want to be Rock 103.5 or the Blaze. Remember, when Q101 came on, the only place in town that was playing any of those records was the Blaze. But it was so "hair" and guy. We believed that the mixing of styles was one of the keys to success, but we needed something common to connect it, and that was women. When we followed them, we always had success in the ratings.

It was as simple as … I kept looking at sales figures and going, "Just look at this."

MARY SHUMINAS

The audience was so forgiving. For them to put up with listening to Don Henley for whether it was three months or six months, they were very forgiving. They really helped evolve the station.

CARLA LEONARDO

People went nuts for the format. It was an incredibly exciting time.

MARY SHUMINAS

Initially, we did well 18-34, but initially it did *real* well 25-54.

CHUCK HILLIER

Ratings shot up, billing followed.

VANCE KORETOS

I had asked the PD and MD many times why we didn't play the new "Seattle sound," and the response was, "We really don't want to play that grunge music." At the time, those bands were considered more hard rock. It wasn't until a year later that Q101 finally played bands like Pearl Jam, Nirvana, and other grunge bands. I remember when Pearl Jam's song "Black" was first on the (listener research) survey. Amazingly, everyone I called voted "like a lot" for the song and the same for the rest of the staff. The song was finally added; it was like the pen was being moved by a higher power.

REY MENA

We were late to understanding the importance of Nirvana or Pearl Jam, or the Seattle movement. The station didn't get that. Programming didn't get that.

CHUCK HILLIER

Because we were a hybrid, we were bowing to occasional perceptions of what we ought to be because of what others were doing in America—at KROQ in Los Angeles, for example. Well, it was a very different radio station from Q101 in Chicago. Gradually, over time, that kind of influence sort of crept in more and more.

BROOKE HUNTER

It was such a huge deal when we first played Pearl Jam.

MARY SHUMINAS

We started playing the conservative stuff, like Pearl Jam's "Black" and the acoustic "Plush" (Stone Temple Pilots).

BROOKE HUNTER

That's when things turned around—Pearl Jam, Nirvana ... Stone Temple Pilots.

MARY SHUMINAS

By the time Pearl Jam's *Vs.* came out, we were definitely all over it.

PASS THE MIC: NEW BLOOD

CHAPTER TWO

As Q101's music direction started to take shape, the station made an effort to bring in air personalities who would complement the new grunge-heavy playlist.

The first major hiring wave included the laid-back Robert Chase, local boy-next-door Steve Fisher, and Brian Peck, a jock with a booming radio voice and a Top 40 pedigree.

A tsunami of on-air talent followed the initial wave, including jocks from high-profile backgrounds (Mark Goodman: MTV; Bobby Skafish: WLUP, WXRT) and absolute obscurity (Brooke Hunter, Samantha James, Brian "the Whipping Boy", and … me).

CHUCK HILLIER, *General Manager*

I think even more interesting was the story of how Robert Murphy came not to be with us.

This guy had a heritage and he was really well known. This was a case where maybe not a lot of research shed light on this topic. There was an undeniable feeling on the part of a couple of people within the corporate campus that Murph was just so ingrained with the pop AC format that he couldn't make the switch over.

CARLA LEONARDO, *Air personality*

Murphy is a great talent, there's no question about that. For where we were going, it was, like, "This guy's getting out of limousines at symphony events." It's not going to work, you know?

ALAN SIMKOWSKI, *Director of Market Development*
You had Robert Murphy walking around in pink suits in the morning. (We were) starting to play the Cure, and then you had all these cool young kids starting to be DJs and playing the Cure. There was a lot of transition going on.

BRIAN PECK, *Air personality*
It was post-format change, but Robert Murphy was still there … I could tell he was on his way out, because we had already started the new format, and I came in as part of that.

ROBERT MURPHY, *Air personality*
The story maybe you want is, as the date got closer, we knew something was going on. For one, they had fired all of my staff: Eleanor Mondale, Dan Walker, and everybody—I was kind of there by myself.

CHUCK HILLIER
I had battled long and hard that there would be a legitimate effort to try to make it work. I felt like (Murphy's) value and advertiser friendliness and revenue from the morning show was worth not just a kneejerk, "I don't think it'll work." People were saying this as though they were betting on a horse.

Regrettably, people above my level made the decision that there would be a change, and like many times after that, I wore the jacket. I got praise when I didn't deserve it; I took some heat when it wasn't my decision. I think Murph could've morphed in reasonably well, particularly at the beginning. It wasn't a modern rock radio station by any stretch. It was adult alternative, and it was a pop alternative.

BILL GAMBLE, *Program Director*
I remember that, as the radio station evolved, and it evolved very, very quickly from what the original research project said, and what we sort of felt would be there. As it evolved quickly, we realized that Murphy wasn't going to be a good fit. And, I think Robert sort of realized it, too. It wasn't exactly what he wanted to do.

ROBERT MURPHY
I went to Bill himself and said, "If you're interested in a buyout with no questions in my contract …," because I had about 10 months left on it. I said, "I

don't have a problem with quietly stepping down." I didn't hear anything about it and then four or five weeks later, I heard I was stepping down. Maybe I was smart enough to realize it was time to go.

BILL GAMBLE
And so, we said, "Well, we have to find a morning show."

CHUCK HILLIER
We parted ways with Murph, and thus began the great hunt to replace those kinds of ratings.

ROBERT MURPHY
All management there treated me with the utmost of professional respect, from the day I was hired to the day I left.

ROBERT CHASE, *Air personality*
I'm a small town guy who grew up listening to WLS, grew up listening to WXRT. I always wanted to go on the radio in Chicago, the market I thought was the greatest radio market in the country. That was always in the back of my mind, to figure out a way to end up there. Haunting Bill Gamble the way I did after leaving Grand Rapids—I mean, I hounded him. He finally cracked and met me. The next thing I know I'm working part time in Chicago, cracking that microphone for the first time late at night. It was the coolest. It was me. If I remember correctly, I was handing off to Robert Murphy and Eleanor Mondale. And you could sense that, you know, we're not done with the shakeup here. John Mellencamp and Steve Winwood were still being played. Certainly, the tide was changing significantly from the results of the whole grunge thing and the exploding alternative scene. It couldn't be ignored—we were kind of dancing around it ….You could just tell there were some pieces and some personalities that—it was going to be interesting to see how it shook out. For me, I was just like, wow, if this lasts a few months—I did it! I was on in Chicago! Hip Hip Hooray for me!

MARY SHUMINAS, *Music Director*
Robert came from Grand Rapids, Michigan … and was just a real music guy. He was, I think, one of the first hires after we were weaning out the adult

contemporary people, so he just seemed so cool – no disrespect to the adult contemporary jocks that had left. He was Mister Alternative … he was kind of looked on as an authority because he was one of our first true "alternative" jocks.

VANCE KORETOS, *Research*

I think that the station started to take shape as a legit "Alt Rock" station when Robert Chase arrived. I think it was his calm and cool delivery, plus his knowledge of music. He was a fan; it was not just a guy pushing buttons for a paycheck. He showed a passion for the music.

BOBBY SKAFISH, *Air personality*

I thought Chase was a great guy, and I remember I shadowed him one show, when I was just signing on to the station. Chase was like, what's wrong with this picture? I should be taking tips from you. You're sitting here watching me run the board?

CARLA LEONARDO

I adore Robert. Chase was able to bring an old school sensibility with an irreverent edge that served him well.

SAMANTHA JAMES, *Air personality*

(Robert Chase) was very laid-back. He was a hippie, but very passionate about the music. I can remember listening to him before I even interviewed. I used to love him and I used to love Carla; they were the two that I would listen to before I even had an interview at Q101.

BRIAN "THE WHIPPING BOY" PARUCH, *Air personality*

Robert Chase is great. I've never really heard anybody say a bad word about him, and deservedly so. He was very unassuming, almost unassuming to a fault. He didn't really seem to have an ego. If he had an ego, it was pretty well hidden, and he was very self-deprecating. Knew a ton about music, would've fit well probably on XRT, and everyone always said that.

BILL GAMBLE

Robert, like, I think, all good music jocks cared about the music. There're a couple kinds of jocks, I think. There are technicians. And I don't mean that in a demeaning way at all. But there are technicians that can be great in almost any format. And then there are jocks—I think (James VanOsdol is) one, I think Chase is one, I think Skafish—that, they really care, or they appear to care about the music they're playing on the radio. I can't judge anybody's true intentions, but they're very believable.

STEVE FISHER, *Air personality*

(Chase is) just a class act. Humble. Intelligent. Articulate. Funny. Dry sense of humor. I mean, just a great guy.

BRIAN PECK

Chase and Steve Fisher and I were kind of, sort of, in that same time, that we came in together.

STEVE FISHER

I grew up listening to what was considered alternative rock at my college radio station …. So, we were playing the Red Hot Chili Peppers back when George Clinton was producing their albums, you know what I mean? We were playing everything from R.E.M. to Depeche Mode to The Cure. Even Metallica was considered cool, because it wasn't like most metal bands. We would play Black Flag; we would play all this cool stuff in college radio. So, I had a love and affinity for alternative music. So, as soon as I heard Q101 flipping, like, this may be cool. And so, when I sent the demo, Mary got it in front of Bill; Bill called me, said, "Hey, we'd like you to audition." So, I said, "Okay." So, I came up; I auditioned. A week later, I got hired to do nights, so. So, I was hired to do nights in November 1992, and the station was starting to take off. And it was still kind of like XRT-lite.

ROBERT CHASE

(Fisher's) the nicest guy. He was a hometown boy, which I leaned on a little bit, because of his knowledge of the avenues, and the streets, and the neighborhoods, and the talk, and the places. I learned a lot from him, just by listening and asking questions. He was, I think, kind of in the same boat—he's arrived, he's doing afternoons in Chicago, and we could kind of see the puzzle coming together, and the music continuing to shift and things developing.

CARLA LEONARDO

(Fisher) was irreverent and sort of becoming the prototype of what would become known as a Q101 jock. He was not old school.

BOBBY SKAFISH

Fisher was a talented guy. He was real good at thinking on his feet; I thought he was really good on the phones. I think he had a quick mind, a facile wit.

CHRISTIAN "CAP" PEDERSEN, *Promotions Assistant*

Steve Fisher. You think of that name and voice and it transports you back to the days of the Q101 Alternative era. You remember him introducing artists like Hole, Blind Mellon, The Breeders, and Soul Asylum. His voice was synonymous with that period of time with alternative music. Steve is genuinely a really kind person and would always take time to talk to listeners at different station events.

ALAN SIMKOWSKI

Steve, we were kind of the same age … he was from Chicago originally—a young guy who could resonate with the audience. The guys liked him; the girls liked him. It was a really good mix, on many levels. He came in at the right time at the right place.

PHIL "TWITCH" GROSCH, *Promotions Assistant*

Steve Fisher was on the air during the period of time that introduced me to Q101 … he was the one I would always hear. He had a very, fun, natural on-air style, and he just, to me, embodied everything that Q101 was. And he was so natural—he was what I aspired to be on the air.

ALAN SIMKOWSKI

You had the old guard meeting the new guard. And we were the cool kids on the block; little did we know how cool we'd eventually be, because that was only the tip of it.

ROBERT CHASE

(I did) a couple months maybe, of doing overnights, and Bill shifted some people around. I think (Steve) Fisher was doing evenings, 7-midnight. He went to afternoons, and I went 7-midnight. It was spring of '93.

BRIAN PECK

It was so confusing, because there were so many changes. Fisher was doing nights, Carla was doing afternoons; then, Fisher was doing afternoons and I was doing mornings. Everybody was switched around all over the place. Chase went to nights, and Carla went to (mid-day).

STEVE FISHER

I just remember the first interview I ever did on Q101 was Dada, and the funny thing is, is that we had to put Dada in another studio, because they couldn't fit in ours. What I did not know at the time was that we were in seven-second delay. And for whatever reason—so, whatever reason, whenever I was trying to interview Dada, live, with no engineer there, just myself trying to figure this whole thing out, I would ask them a question, they would get the question seven seconds later, and then they would answer it. And this went on for about a minute. I'm like, "Okay, something's not working out here." And so, I'll just never forget having that conversation with Bill Gamble, the next day, and he's just pissed. He's just, like, "What happened?" You know how he pounds his fist on the desk. Like, "Steve, we can't have that!" I'm, like, "Bill, nobody told me!" He's, like, "I just,—I don't know why you kept going with it." I go, "'Cause I thought it was kind of funny after a while." I'm going, "I'm thinking for comedy effect, 'cause then I realized what was going on. I was trying to explain it to the audience, and they played along with it. They played their one hit, "Dizz Knee Land." And that stands out just because of what a train wreck it was, truthfully

MARK GOODMAN, *Air personality*

If I remember, I was calling around; I was looking for work. I was out of work, and my daughter was going to be born. I just started calling people and going to people, going, "I'm not working. It's me, Mark Goodman from MTV—do you want to hire me?" I needed work; what are you going to say? I've got to work; I've got a kid coming.

BILL GAMBLE

I knew Mark Goodman a little bit from Philadelphia, when he used to be a DJ back then, somewhere in the '80s, before he went to MTV. And he had been working, I think, in Los Angeles, but he was out of work in LA or maybe he was doing part-time somewhere in LA. And we thought that Goodman, and he's got some MTV credibility, and you know, on paper, it sounded like a great plan. But it was the start of the string of morning shows that didn't work.

ALAN SIMKOWSKI

The Mark Goodman thing, I thought, was, for a while, really cool. I mean, here was a guy that everybody idolized on MTV. He hosted Live Aid, which I think is still the most significant music in my lifetime.

HEIDI HESS, *Air personality*

He was the first VJ on MTV, from the first video jocks. Being around Mark Goodman was a treat.

SAMANTHA JAMES

I liked Mark. I, of course, was a little star-struck when I first met him because I was like 13, 14 years old watching him on MTV. I thought that was really cool.

BRIAN PECK

Mark was a bit of a musicologist. He knew the artists; he had his connections with everybody.

MARK GOODMAN

One of the great interviews that I've ever had happened there with Tori Amos. It was right around the Olympics when Nancy Kerrigan got her knee smashed. Everybody was talking about that. It was like the hugest worldwide news. She was really philosophical about it and was just fascinated by how it just represented who we were. It was sort of classic American stuff—the characters involved, Nancy Kerrigan, this trailer park chick, and her freakin' loser boyfriend. What a story-- she was fascinated by, from the story perspective, what it was. I bought her act; I thought that she was real. I thought she was very genuine. I enjoyed meeting her.

BROOKE HUNTER, *Producer, Air personality*

I'd watched him on MTV back in the early 80s when MTV launched, but that kind of went away after a few months, and then, he was just another person I was trying to prep and get ready for a morning show. Very challenging prep, at that ... I was trying to provide him with information, but at the same time he thought—and rightfully so at that point—he knew more than I did, so...

MARK GOODMAN

I rarely saw people except for Brooke; that's the person that I remember working with the most. At like one o' clock, I'd be out of there, falling asleep. I never really felt totally, 100%, physically there. That shift just killed me; it ruined my life.

BROOKE HUNTER

He worked with me and he wasn't uncool; he wasn't a nasty person to me necessarily, but I don't think that he treated me like a regular producer who had the experience. I was gaining the experience as I was working for him.

MARK GOODMAN

We called up to Liz Phair's house one morning—Brooke Hunter got this, she somehow got her number, so we called and said, "Is Liz there?" Her mom answers the phone—she's living with her mom. Her mom answers the phone—"Is Liz there?" "Yeah, she's here, but she can't come to the phone right now—she's in the shower." I'm like, "Really? You've got to go in there! We want to talk to Liz Phair in the shower."

BROOKE HUNTER

I remember doing a Lemonheads performance at the old Avalon club. That was when Lemonheads were hot with "It's a Shame about Ray". That might have been one of our first productions. I went, and Mark introed the band, and it was fun --that was one of our very first listener-involved events.

BRIAN PECK

I always thought Mark was a great jock. I always thought he did a really good job. Interviews, he was really smooth; he was really good.

SAMANTHA JAMES

I settled in Chicago because my aunt and uncle lived in Naperville. I basically wrote Bill Gamble a letter–in fact, I don't think I even wrote it to Bill Gamble; I wasn't even experienced enough to know who to send it to. I'd been listening to the station, and I loved the station because, at the time, it was the Smiths, the Cure, and Depeche Mode, just fantastic music from my perspective. I said, "I love your station, I'd love to come in, I'll make the tea, I'll answer the phones, I'll type letters, whatever you need, I'd just love to come and work for your radio station." I threw this demo tape in, just for a laugh, and then, I got that phone call. I came in wearing a business suit, because I didn't know what I was interviewing for—I didn't know I was interviewing for a DJ; I thought I was interviewing for a secretary or a receptionist or something. And Robert Chase can tell you, I lied through the interview, basically saying, "Oh yeah, I used to be on the air in Radio Luxembourg." They put me on the air for my audition with Robert Chase, and I could tell you, he knew I'd never been on the air in my life because I didn't know how to work the board at all. I didn't have a clue. He was going, "Well, you need to open the mic, and you need to have your headphones on." Thank God, I got Robert as the person I got the audition with, because I could've had somebody that wasn't quite as welcoming. That was my story, the whole thing—I was just so lucky.

BILL GAMBLE

I didn't know that. I was a sucker for the accent. Good for her. But you know, that's —those are the kind of people that succeed. You know, because they, I guess—probably wrong adjective—but they have the balls just to do it. And, again, it goes back to being fearless. "Oh, what if I get this job? Oh, my gosh, I have to be on the radio and I've never done this before!" Okay. Didn't seem to bother her, seemed to work out fine.

SAMANTHA JAMES

(Bill) called me and said that I had the gig, but he wanted to know what name I wanted to use on the air. My name at the time, my maiden name, was Tuck. He obviously knew that probably wasn't going to be a great radio name, Samantha Tuck. He asked me what name I used on the air, because, of course, I'd made out like I'd been on the air before. And I had been babysitting that morning—a little boy called James, and I completely panicked and I was like, "Uh … Samantha James." And that's how I came up with the name. I was completely flabbergasted. Even when I had the audition, I thought, well, that was fun. I was on the radio for a little while.

BOBBY SKAFISH

I have certain songs burned into my mind from when I would play them about twice a shift when I was doing the overnight show. If I hear these songs, it takes me right back: "Into Your Arms" by the Lemonheads, "Linger" by the Cranberries, "Devil With the Green Eyes" by Matthew Sweet, "Low " by Cracker, "Connected" by Stereo MCs. There were a few songs from the *Coneheads* soundtrack: "It's a Free World Baby" by R.E.M. and "Soul to Squeeze" by Red Hot Chili Peppers. If I hear the beginning of "Linger" by the Cranberries, just the way it kind of sneaks up, it's three in the morning, my skin looks kind of gray/green, and I'm working the controls at Q101. It just all comes back to me.

MARK GOODMAN

That Thanksgiving (morning show concert broadcast at China Club) was a fantastic thing; I was really proud of that. The reason that it came up was not for me to hype the morning show; it was because I wanted to do something that was great around Thanksgiving for a cause I believed in. We did some research on it, and there were things that I thought made sense, and things I thought I could be passionate about, and the station supported that. It was great, and I thought it led to a great moment.

ALAN SIMKOWSKI

That breakfast broadcast was the starting point of all those different types of live events. The China Club was this legendary club. I still remember the bands – Matthew Sweet, Material Issue, Catherine Wheel, and the Indians. People started waking up on that.

BILL GAMBLE

Yeah, that was the first show, and, we thought it was going to be okay, but we decided that … we'd start small, and see how it went. None of us, at least I, didn't have any idea how it was going to do until I rolled up to the China Club, and it was, I don't know, 3:30, quarter to four in the morning, 'cause we were going on the air at 5. And there were all these kids just lined up outside. And it's just, it's one of those moments I'll always remember, 'cause I'm thinking "Whoa! This is really going to be big!"

BOBBY SKAFISH

The headliner was Matthew Sweet, but the act that went on directly before was Material Issue. Material Issue was really out for blood. I think they bested Matthew Sweet and company on stage, because I think they were hungrier. Jim Ellison really worked the room, really went after it. I'm not going to use the phrase "blew them off the stage," because that's a little extreme. But I think that they were hungrier, and I thought they went over better. It was a very young crowd, because the station had such a young appeal. It reminded me—I remember saying this to Chuck Hillier—that this reminded me of some teen event WLS would stage in the '60s: you know, hey, here's the New Colony Six and the Cryan' Shames, and all the teenagers would get off on it. It had that kind of feverish energy. That's a really nice memory.

BROOKE HUNTER

We did the broadcast at China Club and Chuck, at the time, had never seen or heard of a mosh pit. So all these kids start jumping in, and like, throwing each other around on the floor, and Chuck was freaking out. Chuck was like, "You've got to get those kids out of there. What are they doing? They're going to kill each other." Bill is like, "That's a mosh pit. You cannot stop them from doing that." So like, the very first time that we ever had something where the audience was really involved and was really getting into it and started, like, throwing each other around was down on the floor of the China Club.

MARY SHUMINAS

I'm going to guess it was during Material Issue's set, the kids in the audience were crowd surfing. Our General Manager, I don't believe he was up on that concept. When he saw that, he was like, "We've got to do something! Someone has to stop this! What are we going to do?" Obviously, there was no stopping it. It was eye opening for Mr. Hillier, I guess.

BILL GAMBLE

He goes, "There's a mosh pit going on on the floor, and people are going up." He goes, "You know, someone's going to get hurt! You've got to make them stop!" And I go, "What? How would I be able to make them stop? It's not like there's a light switch, and (you say) 'Okay kids, time to go home, the slumber party's over!'" Yeah, well, it was classic, and it was, I think it was just … kids were doing what they thought they were supposed to be doing.

MARK GOODMAN

That was a push for the morning show; that was one of the rare moments that the station did get behind me to do something. They didn't do any work, they didn't promote. I didn't see any fucking billboards for Mark Goodman. I didn't see anything like that. There was no real commitment to it. The kind of show that I would want to do in the morning was not the morning-zoo-type bullshit. I just wasn't really into what Bill Gamble wanted me to sell. In terms of a cool factor, I didn't think that Bill had any fucking idea. He could've been programming polkas.

BRIAN PECK

When we came back from (broadcasting from the Olympics in) Lillehammer, I remember Gamble called us in. And he told me, "I feel horrible." He just fired Mark right before me that same day. "You and Mark had never sounded better than when you were in Lillehammer, but I'm blowing it up. Since I fired him, I've got to fire you. So here's some severance," and he gave me a nice big fat check. He says, "And I'm going to rehire you this afternoon. And you're going to be my fill-in guy, and you're going to do all the appearances, and I'm going to make sure you make just as much, if not more, than you were making doing the morning show with Mark." And doggone if he didn't. He came through on his word. I was doing five appearances a week, making a ton of dough in cash, and then filling in for everybody.

HEIDI HESS

Mark was so nice. He dressed great; he was cool. He was upbeat. I think he was a really nice man. I think he gave it his all. I think that he was behind the mic, so it wasn't Mark Goodman on television anymore, and people had to relate to him on the mic and that might not have been the easiest thing.

MARK GOODMAN

He was really—although (Bill) wanted to play music, I think he wanted to play music because he wasn't really sure about how I was going to be as a morning personality. Ultimately, that was what he didn't believe in after hearing me for a year. He actually said the words to me, "I just don't hear you in mornings." And I have to agree—I'm not the guy who did the best morning show ever there. I put my heart into it, and I think we did some cool stuff, and I think we had some great interviews, and played some decent music—I would've liked to be more adventurous, but that's just the way a lot of us are.

BROOKE HUNTER

Bill told me that Mark was no longer with the station. But that I was okay. They would need me to produce for the next morning show, whoever it was; I think, at that point, it was Bobby Skafish.

BOBBY SKAFISH

There was a whole succession of morning shows that got very brief runs at it. It was one day out of the blue; I wasn't expecting it. Bill called me into the office, told me he was making a change, and would I do the morning show? It would still be a six-day week, which didn't thrill me. The money was not very impressive either, but he made it sound like, "Let's try it out, see how it fits, and then, we'll see down the road."

SAMANTHA JAMES

Because I hadn't lived in Chicago for very long, I didn't know the legend that was Bobby Skafish. When he came to the station, everybody was like, Bobby Skafish! Bobby Skafish!

HEIDI HESS

Bobby Skafish is the coolest cat … He's somebody that can make something new out of an everyday PSA. That guy can go on the air, open the mic, and say things; he can make something sound so interesting, when it's not, and also find different ways of speaking about the same thing that we all speak about. No one else does that.

SUSAN BANACH, *Air personality*

Bobby Skafish was really sweet. He always had some really good stories. He actually talks the way he talks on the air. He's actually, "Hey man," hippie, cool, using all these beatnik terms—what you see is what you get, and that's who he is. He's not putting on any airs when he's on the air.

BROOKE HUNTER

When I produced for Skafish, that's when I went to Bill and said I thought I could be on the air. So, he hired me for part-time work but I was still doing the morning show producing; so, a lot of times I would do the overnight shift and stay in the studio to produce for (Bobby). So I was literally in the studio for like 10 hours. That made (Bobby) really angry. There were a couple times, literally, I was so tired I'd fall asleep while he was on the air.

BOBBY SKAFISH

When I first began, they assigned Brooke to be my producer. But simultaneously with that, Brooke was being broken in on doing shows that she would host, so she was kind of in the process of graduating from being a morning show producer to being a host.

MARY SHUMINAS

Brooke, she actually started in the research department. She was very enthusiastic and loved the music. She was in the demographic, and I think just asked Bill for the opportunity to get on the air, and he auditioned her in overnights, and eventually, she worked almost every day part at that radio station.

CARLA LEONARDO

Brooke, she's great, she's great. It was really cool because she, and (James VanOsdol), to an extent, pretty much lived a certain type of dream, really. You're working as interns, or working in the research room, and they pluck you out; they see you, and bring you in to be part of this radio station. I think that's terrific. It doesn't happen every day, and it certainly doesn't happen now; so, it was a really exciting time to watch young talent develop.

STEVE FISHER

I think what's great is that Gamble really broke the mold. Like, I mean, to hire people that started off as interns, and just throw them on the air. To give Whip and Brooke (and James VanOsdol), and just—I think it's great because Gamble's philosophy was simple. It was just, like, "Hey. They've got a good personality. You can teach them how to do good radio," which is just the opposite of what most program directors sort of think, like, "Which call letters did you work for, and what were your ratings? And let me hear your demo." It was so the opposite that he got it. It was like, "Look, we need to hire real people." People don't want to be fooled by jive, as Bill would call it; people don't want to be—they know BS, especially the alternative audience.

BRIAN "THE WHIPPING BOY" PARUCH

I always felt like (Brooke) really took everything the right amount of seriously, which was not very. I don't think she was ever really very stressed about anything; I think she took it all like, "Eh, whatever, it's radio," which is kind of like the best attitude to have.

KEITH SGARIGLIA, *Promotions Assistant*

I felt like she was the rebel of the staff. She kind of had that attitude, like, I'm doing whatever the hell I want to.

ERIC "SHARK" OLSON, *Air personality*

She had a terrific "WTF are we doing?" energy, which, to me, helped (us) shape our "stationality" and take some of the risks we did. Like most of us, Brooke really wanted the station to push itself and be true to the alternative label.

BRIAN "THE WHIPPING BOY" PARUCH

She had a little bit of an edge to her, obviously, and I think that's what Bill first saw in her to put her on the air.

STEVE FISHER

She's like a sister. I mean, we really—we always had—there was a time when we didn't get along great, but I think over the years, having become parents, we've gotten to know each other. She is real. She is authentic. She is raw. I think that's what you want in a radio personality.

BROOKE HUNTER

James VanOsdol, Brian "the Whipping Boy", and I all kind of started off at the same time.

MARY SHUMINAS

Brian worked his way up from an internship. He was actually working at WPGU in Champaign, and Brian expressed an interest to Bill in going on air. Bill would have Brian send me airchecks (recordings of on-air talk breaks), and I would critique them. Eventually Bill gave him a shot, as far as auditioning, and he started doing weekends. He eventually did sports, and news, and traffic. Brian was very versatile, just a smart guy, and was able to handle anything on the station.

BRIAN "THE WHIPPING BOY" PARUCH

I seriously think (Bill) probably needed a warm body that was even competent. I guess a lot of radio stations, even ones who aren't trying to be, like, new, cutting edge and young … aren't afraid to put some kid from the office on in the middle of the night on a weekend if they're willing to work seven-hour shifts.

I'm sure his thought process was, I was young, kind of like the audience in a way, and … sounded okay enough. What he told me was, "As long as you don't say the 'F word' and you play all the commercials, you'll be fine."

ROBERT CHASE

Brian, the Whipping Boy. I trained him. Great kid, right off the bat. A sweetheart. (He was) well educated, and was bright-eyed and was excited to have an opportunity.

BRIAN "THE WHIPPING BOY" PARUCH

Robert Chase, who was doing nights at the time when I first started, he trained. So, I came in some night and was supposed to spend an hour with him learning the board, or whatever. First thing he did, "All right, you know what you're going to do? You're going to intro the next record. I'm going down the hallway to make some coffee. And I'll see you after you do the intro."

ROBERT CHASE

I was out in the hall, and nothing happened and the song started. I don't know why he didn't turn the microphone on, but it didn't happen, and that was the only time for him. He was a quick study, a quick learner. Brian just took off from there.

BRIAN "THE WHIPPING BOY" PARUCH

He just left me alone in there. And I hadn't trained on the board or anything. It was basically, "Hit this button, turn the mic on, talk, then turn it off, and I'll be right back." When I turned up the mic, nothing happened, and so, I had dead air. Then I just hit the song, and I was like, "I don't know what happened." He was like, "Well, did you turn the mic on?" I'm like, "Oh, you have to turn it on?" because the mic on the board in Champaign, you didn't have to actually press 'on', you just turned it up and it turned on for you.

To this day, I can't remember if it was Henry Rollins' "Liar" into Sarah McLachlan's "Possession," or vice versa, but it was definitely those two songs. That was Q101 then, right? By the way, on a side note, I made fun of that segue (on the night of Q101's final broadcast in 2011) on the air, and one of the guys was like, "Yeah, you weren't here for 'Shuffle.' That's not that weird."

TOM KOKINAKOS, *Research*

It was such a fun environment. I remember when I was at Columbia College and I set my goal as wanting to work at Q101. It was the station I listened to all the time. I couldn't believe, at the time, that I actually worked there. My friends were jealous and would be like, "Is Steve Fisher as cool as he sounds?" and, "Brooke sounds really hot; is she?" I was never treated as just some college student or part-timer, except I did have to go get lunch for everyone; but hey, doesn't every intern?

ZOLTAR, *Air personality*

I remember very well when I got (to Q101) and did the audition. Brian Peck was the one that gave me a quick rundown on all the equipment and everything, and he sat in with me on my first night to make sure everything was going cool. I remember saying to Bill Gamble, "Look, no disrespect, I want the job, but you've got a guy here who's pretty fucking good." Gamble said, "We're missing something on the station; I need a station freak. I need somebody that just does things on a whim, a little on the wild side; that's why we want you."

BRIAN PECK

Zoltar, we brought him in to do overnights from WHFS in D.C., which was a really cool alternative station. Zoltar really knew the music.

ZOLTAR

The first night, when I was doing my audition, was when (Robert Chase) was doing the 7-midnight. I remember, it was my last audition, and we would do these really wild changeovers. I said, "This is my last night for auditioning." I said, "Hopefully Bill Gamble will like me." He said, "Yeah, he's going to hire you. I've got a good feeling about you. "

BRIAN "THE WHIPPING BOY" PARUCH

He'd been in San Diego, and Tijuana, and the Navy SEALs, or whatever.

ZOLTAR

It was Skafish and Carla Leonardo in the morning; Samantha was doing mid-days; Fisher was doing afternoon drive. Robert Chase was doing the evenings, and I would do the overnights.

MARY SHUMINAS

Zoltar did the overnight show, called the Industrial Zone. And he played a lot of hardcore club music and actually did his show live. I think we did a lot of tie-ins with clubs. He played, on the mainstream level, Nine Inch Nails, Nitzer Ebb, Skinny Puppy, the more hardcore club music.

TOM KOKINAKOS

Zoltar was the man that got me hooked on industrial music and bands like Laibach, Pigface, and Lords of Acid.

ZOLTAR

When they finally gave me the Industrial Zone, which was a few months after I was hired, (Bill Gamble) goes, "This is about station image. It's not about numbers; this is about image." He goes, "I want you to portray a certain image." Remember how many things were mandatory - things we'd always have to do or places we'd always have to go? Bill Gamble would call me up and say, "Just don't show up." If they ever go, "Well Bill Gamble said you're supposed to show up," just say "So?" There were a lot of things that I would just go "Pffft, whatever." He loved that stuff. I was like the anti-guy.

BRIAN PECK

I would hear all of Zoltar's stories. I would witness him with his—how do I put this?—the way he would meet women. He was not afraid of anything. He met so many different women, and beautiful women. We would go places, different clubs. We would hang out at Baja Beach Club on Industrial Night. That's when all the freaks came out, and Zoltar loved it there. And I had appearances there every night through Q101, so he would come out every week and we'd hang out. He was just a really good guy. I knew him as Jeff, Jeff Gibson, just this really good-hearted person - really cool. He was extremely interesting.

MIKE BRATTON, *Creative Director*

I loved my job. I loved hanging out late at night. I loved to see the really, really, crazy hot, weird Goth girls Zoltar would bring into the station in the middle of the night.

ZOLTAR

Chicago, it was, like, if you were on the radio, you were like a rock star. And that's how I got treated.

ALAN SIMKOWSKI

If you were a DJ at Q101, you were a rock star and you were getting paid big bank to go do appearances.

MIKE BRATTON

I'm pretty sure that Zoltar would light candles in the studio when he was in there at night … also, he was a giant of a dude; he was, like, six feet tall, 5'11, somewhere in that neighborhood. He wasn't, like, super tall, but he was huge, and built like a Navy SEAL. Whether or not he was a Navy SEAL, I'm not 100% sure, but most people thought he was, so I was willing to go along with that story.

KEITH SGARIGLIA

There were candles all over the studio. He just had this—I think some of it was legit. If I remember right, he was a Navy SEAL. He was a badass. He was in really good shape. And he would go out to these clubs, and people would just love him. His audience was just obsessed with him.

BRIAN "THE WHIPPING BOY" PARUCH

Zoltar officially got in trouble, or yelled at or 'memoed' or talked to or whatever, for burning candles in the studio and dripping wax into the equipment … the candles were every night; that was his thing.

ZOLTAR

I always wanted to set the mood; I wanted to set the vibe. It's one of those things where, yeah, you can say you're doing things, but if you try to live it, you can make it more sellable. And to me it really wasn't a stretch. With the things I did on air, it really wasn't a stretch, you know? And that's what made the whole Q101 experience, for me, the best radio experience I've ever had. By far. By. Far. Simply because there was that trust—Bill Gamble, for some reason, he trusted me to do the right thing.

BILL GAMBLE

I sort of have a rule: if I don't see it or know about it, it doesn't exist. It's a style of management that I guess served me okay. But yeah. You have to remember. I mean, I know there are industrial music fans, and that's why Q101 streamed early, because we knew, worldwide, there might even be an audience. But the reality is, is it's like, it's not the most listenable thing that you can put on the radio station. I mean, that was all done, it was done after one in the morning, when the ratings stopped back then, and it was all done for image. We had to put something on that was cooler than the radio station, and that was Zoltar. So, we're playing "Hand in My Pocket," and he's playing KMFDM. It was about making sure the brand still was cutting edge. And you couldn't be that cutting edge during the day.

ZOLTAR

A lot of times, I would, like, say, "Hey, let's check out some new music." And I would get something in the mail or something like that, and I would just throw it on. And I remember there was this one song, this woman who kept going, "Fuck you, fuck you, I hate you, I hate you. Fuck you, fuck you, I hate you, I hate you." And after about 10 times of doing this, I cut the song off and I go, "Okay, we got the point." Bill never said anything to me about that.

BROOKE HUNTER

A lot of people thought, at the time, "Oh, you and Zoltar are doing each other." Zoltar was like my brother. I mean, to this day, I still talk to Zoltar. I still laugh, because James VanOsdol always said, "I think Zoltar's going home and watching *Friends* every night. I don't think he's really as crazy as (he seems)."

ZOLTAR

Oh, Brooke was just, like … the one woman at the radio station who thought a lot like I did. Brooke was my buddy. I knew that I could say anything to Brooke and it would never go anywhere else. Never.

MADISON, *Air personality*

I think he was wild, and perverted, and sexual, and crazy, and he was a super nice guy, too.

ZOLTAR

I had this book, this S&M "how to" book and I lent it to Brooke. She was talking to Fisher one day on the air. She goes, "I've got a blind date tonight, but you know what? Zoltar loaned me his S&M book; I'm kicking somebody's ass tonight!"

ALAN SIMKOWSKI

(Zoltar?) One of the coolest cats in the world. Loved him. Loved the music. He was just a cool cat.

ERIC "SHARK" OLSON

The fact we had Zoltar doing something awesome and different late night really helped us keep an edge and a true connection to fans of the industrial brand. Other than some Stabbing Westward or Gravity Kills, we didn't play much of the heavier stuff on the station. But when Z showed up at midnight with CDs in his rubber-spiked backpack, he brought something very different, unique, to overnights and had a huge audience and fan base.

ZOLTAR

I knew what (Bill) would let me get away with, and I knew what I couldn't get away with. So, I just kept it up at the line, never crossing it. And people used to go, "Dude, did you hear that song?" "Oh, yeah, I'm playing it." They'd go, "They said 'fuck' in that song." And I'd go, "… are you sure?" Or I would just deny it, or I would just go, "Yeah, and …?"

SAMANTHA JAMES

He was one of those people that were just so nice, but at the same time, so 'effing' cool. He was just so amazingly cool. I loved what he did. He was just so talented, with the music he played - how he put it together, his knowledge of the music. He was just so submerged in it all. He was more submerged in his radio show, I would say, than probably any of the other people who worked there at the time. He was really, really, into it.

ZOLTAR

I had this woman approach me. She said, "You turned me on to so much cool music culture," and she just hugged me and said, "Thank you." I never looked at it like that. I just looked as it like I was just goofing off, having a good time in my room, and mom and dad were sleeping, and they weren't going to wake up and say, "Hey, turn that down." That was my whole thing. That was my whole message I was trying to get across—I don't care what you grew up listening to, or what you're used to listening to, but there's a whole realm, a whole subgenre, of music that's really cool, and you need to check this out.

THE MAN WHO SOLD THE WORLD: THE DEATH OF KURT COBAIN

CHAPTER THREE

The death of Nirvana front man Kurt Cobain on April 5, 1994 didn't become public knowledge until a few days after the fact.

As I type this, I have no doubt that if Cobain had died in 2012, news would have ripped through Twitter and Facebook with time-and-space-warping speed. In 1994, however, radio was one of the primary ways consumers got their news and information. Because of that, when news broke that the face of the grunge revolution and the alternative nation had checked out, Q101 was the first place where many Chicagoans heard the news.

STEVE FISHER, *Air personality*

I think the one defining moment for me, of what I think Q101 could've been more of, is April 8, 1994. And that's when they found Kurt Cobain. And I just remember getting, I mean, Bill just goes, "Put the listeners on the air. Talk to them. This is huge. You have to make the announcement. Talk to them."

BROOKE HUNTER, *Air personality*

I remember *exactly* where I was and what I was doing when I found out he was dead. I was in my little red Mazda Miata, in the express lanes of the Kennedy, on my way to see my accountant to get my taxes done when my phone rang. It was James VanOsdol; James said to me, "He's dead," and instantly I knew that he was referring to Kurt Cobain. It was definitely sad, but not shocking at all.

BRIAN "THE WHIPPING BOY" PARUCH, *Air personality*

I was still in Champaign when Cobain killed himself. I had interned at Q101 the previous summer, '93; then I came back in the summer of '94.

I was actually on the air at WPGU in Champaign when the news came over that it had happened. I remember thinking selfishly that this would be a moment I'd always remember, and that I'd be able to tell my grandchildren about it, because I was the guy who told Central Illinois about the death of a famous rock star. Now, maybe I'm jaded, but I don't think my grandchildren will probably give a crap ... and the specifics of what I said, or whether we took live calls, have kind of escaped me. I do know that when he had *almost* died a few months before that—in Rome, I think—a couple of us at the station said that if he had died, he would have become some sort of mythical, god-like, figure in music history, whose role in our lives would probably end up being exaggerated ... even though they were great and all that. And I believe that's exactly what happened.

BROOKE HUNTER

I was fortunate enough to see Nirvana at the Aragon earlier that year for their *In Utero* show on the night that they didn't have to drag a wasted Kurt off the stage. I remember watching him perform thinking, "Wow, this is the guy who everyone is talking about, who is so 'F'd' up," but on that particular night, they were spot on.

STEVE FISHER

So, I made the announcement. Got flooded with faxes. People were angry; people were hurt. A lot of, as you can imagine, really distasteful jokes, were coming through. People were calling up—people that, later, wrote three page letters, explaining how that day meant a lot to them. Because I think it gave people a chance to grieve, and respond, I thought. You know this was before Facebook; this was before Twitter. Q101 had become a social media. A community. And that was huge. And that's where I think if we had been more like *that*, on a daily basis, people would've been like, "Oh, we don't hate you. We love you. In fact, you are us. You speak to us." And I think that that's where being on air, me personally—I wish I had more days like that, where I really reached out to the community and said, "Okay, here's what we're all about. We're in this together. It's you and me. I feel your pain, having had friends that committed suicide, or having known people that have gone through depression. I've never been a heroin addict and I can't imagine what *that's* like - the state of mind that you're in then. I just think that everybody had to vent, and that was the most powerful day. And I will never forget, like, I literally had friends come up to me and go—I know it was tragic, but they go—"But that was the best day of radio you ever had." And I go, "Okay." And you don't want that to happen every day, so you have to respond that way, but I think the point is that I think that national tragedies could be handled so differently. And I call that a national tragedy because he was the face of alternative music. I mean, it changed everything, I think, for us as a station. I mean, it just changed these kids' perspectives. I mean, for the older generation, it was like when Jimi Hendrix and Janis Joplin and Jim Morrison died, all at the age of 27.

ROBERT CHASE, *Air personality*

That whole day was a bit of a trip, in that …it was a heavy day. That station owed everything to one record, if you think about it: *Nevermind*, and the cultural change that followed, and the doors that it opened for good or bad. To have Kurt 'off himself' was shocking and upsetting. I had met Kurt on the *Bleach* tour in Denver. I had been given an early cassette of *Nevermind* and John Rosenfelder at DGC, a guy I had worked with years earlier had asked me to pick a single. I picked "Come as You Are," which, at that point in time, seemed like a safe song to go with, a really good song. Obviously "Smells like Teen Spirit" had something packing, but I didn't know if radio was ready for it. I remember playing that record for my staff at (W)LAV thinking, this is four songs deep, easily, and most of them in the room were not getting it.

Bill (Gamble) called and wanted me to go down and represent the station on WGN TV with (musician) Chris Holmes. We did that TV special. That was a little unnerving. I was pretty nervous going on WGN-TV, talking about something that was still really hard to get a grip on. That night, (I had) to go on the air and field phone call after phone call, from 7-midnight, from kids who could relate to what he did. I remember taking those calls from kids who were obviously in their bedrooms, obviously trying to work through puberty, thinking that it wasn't a bad choice on his part.

I remember going home that night and talking to (my wife) about how many kids were going to copycat. Having those thoughts, and even that the thought came into my head, was disturbing.

STEVE FISHER

We literally broke down the walls and said, "We're going to get real here." And people were crying; people were pissed. People were … saying tasteless jokes. But, you know, we tried to get everybody's reaction. Some people hated me for saying, "I can't believe he did this."

TWISTED CHRISTMAS

CHAPTER FOUR

On December 2, 1994, Q101 hosted its first annual "Twisted" concert (called "Twisted Christmas" for that first year only). "Radio shows" were quickly becoming a major source of NTR (Non-Traditional Revenue) across the industry, and the alternative format in particular was taking the lead on producing these events.

Q101's first effort lined up a diverse list of talent, including future stars Veruca Salt and Weezer, veteran acts Bad Religion, Dinosaur Jr., and Killing Joke, and the of-the-moment, Courtney Love-led, Hole.

MARY SHUMINAS. *Music Director*

I named Twisted. Well, I stole the name from … there was a comedy album of parody Christmas songs (by Bob Rivers). I remember having the CD, and I borrowed that name. At the time, I looked into it; there didn't appear to be a trademark on the name. Eventually I think Bob Rivers did trademark the name because we got a "cease and desist" from Bob Rivers' attorney or whatever, and we eventually shortened the name from Twisted Christmas to Twisted. We probably could have trademarked the name ourselves, but we didn't bother to do it.

BILL GAMBLE, *Program Director*

We thought that we could put together a show, that we could get some bands (together where) the listeners could see this music. There really wasn't anything like it out there, except some of the old '50s and '60s stars like Tommy James, the same people that are still touring right now, the cavalcade of oldies acts. And we thought, "Well, wait, if we could have it make sense …" And we saw L.A. was doing a show or two, and we thought, "We could probably try to pull that off."

MARY SHUMINAS

The Twisted show was at U.I.C. (Pavilion). This was before radio shows were kind of package deals. Eventually bands started doing these as package deals, and they would do 15 or 20 cities, but at the time, it was a fairly new thing. It was exciting. We did it for imaging for the radio station, and I think there was a charity tie-in to it. I guess, eventually, it became non-traditional revenue; it was a big money-making thing for the station.

That was one of the major motivations behind putting those shows together. But in the very beginning, it was more about imaging the station. It was cool; it was exciting.

ROBERT CHASE, *Air personality*

The first Twisted was at the U.I.C. Pavilion, (with a) great, fun lineup. We had Killing Joke; that was cool. We had Hole, Bad Religion. I introduced Courtney Love—and I remember she said something nice about me on stage. She had noticed me in some way, so that was a bit thrilling.

BROOKE HUNTER, *Air personality*

I was hosting nights, and we did a remote from literally a tiny closet inside the Pavilion.

ROBERT CHASE

Backstage, I remember Courtney crashing Brooke's live broadcast while talking to the Veruca Salt girls.

BROOKE HUNTER

That was the night that Courtney Love came in drunk and I got the very first post-Kurt Cobain suicide interview with Courtney that night. It was insane. It was absolutely insane. She was about three sheets to the wind … she gutted herself.

ROBERT CHASE

That was probably the apex of excitement that night—not knowing what was going to happen next, (with) Courtney bounding about.

BROOKE HUNTER

So when she was getting ready to go on stage, I went in with her to her dressing room, and that was the year—I still have it somewhere—there was a *Rolling Stone* interview that came out with Courtney. She was on the cover. I remember the writing was in a metallic blue and I had a metallic gray Sharpie and she signed my *Rolling Stone*. She wrote on there, "Brooke, please protect me." Like how am I supposed to protect you? I'm just a stupid night jock in Chicago.

Somebody told me afterwards that she hadn't talked to anyone since Kurt had died. He died in April and Twisted was sometime in early December, and she hadn't spoken to anyone or done any interviews up until then, and that was kind of an accidental interview.

REY MENA, *Marketing Director*

The first Twisted Christmas, when we did the concert at the U.I.C. Pavilion, you could just see the reaction of the crowd. As an AC station, you didn't see any of that. You could feel the buzz around the station.

TOM KOKINAKOS, *Research*

Mary Shuminas, who was the Music Director at the time, asked me if I wanted to dress up as the Grinch and dance in between band sets. Whenever I hit the stage to dance, the audience would throw stuff at me and I got really pissed, so I started to flip them off. After exiting the stage, our Program Director (Bill Gamble) said, "Good job, but you can cut out the middle finger."

Right before Bad Religion was to go onstage, Greg Graffin, the lead singer, came up to me and said he was going to tackle me onstage. So right when they came on and I was getting off, he tackled me on stage, stole my Grinch head, and wore it on stage for the beginning of their set. That was totally cool.

ROBERT CHASE

It was a long day, and it was a long night, and it was a feeling of success—we pulled it off. We sold the tickets, and we had some cool bands that played well. (We wondered if) this was going to be an annual thing, and we decided, Yeah!

SATURATED IN GUYVILLE: THE LOCAL MUSIC SHOWCASE

CHAPTER FIVE

For the listeners who were serious—I mean, *really serious*—about music, Q101 wasn't worth listening to when the rest of the audience was plugged in to the station. The real value of modern rock programming, as far as the ultra serious fans were concerned, was in its "specialty programming."

Specialty shows flesh out the overall sound of a station; they are frequently programmed by the people hosting them, and designed to reach impassioned niche audiences who couldn't give less of a shit about the heavy rotation hits of the day. Because these shows fall far enough outside the mainstream, they air at times when the mainstream isn't paying attention. Most commonly, these shows are on Sunday nights. Traditionally, the weakest nights for rock radio listening, Sunday nights are dumping grounds for bartered and fringe shows that generally run counter to whatever music and image goals Program Directors have for their stations.

Q101 has aired a handful of notable specialty shows over its 19-year history. *Electronic Trip*, hosted by local club DJ, Tom Pazen, started in 1997. The three-hour show ran at two in the morning on Sundays, super-serving Chicago's rave and glow stick cultures. *Electronic Trip* would evolve over the years, memorably turning into the Saturday night staple *Sonic Boom*, which hit its peak when Joey "Just Joey" Swanson manned the turntables.

In the 00s, Q101's most risk-taking programming occurred on the weekends, with the interview-focused *On the Bus with Ryan Manno*, and a new music-focused *TBA*, with Robert English and Nikki Chuminatto anchoring.

Early into its history, Q101 put *Sound Opinions*, the self-described "World's Only Rock and Roll Talk Show," on the air. Two of Chicago's most well known music critics, Jim DeRogatis (then of the *Chicago Sun-Times*) and Bill Wyman (formerly of the *Chicago Reader*), hosted the two-hour program.

Now syndicated across the country through NPR, and hosted by the team of DeRogatis and *Chicago Tribune* critic Greg Kot, *Sound Opinions* was totally embedded in Chicago culture when it first appeared on Q101. DeRogatis and Wyman knew the music, the local players, the clubs, and the strengths and shortcomings of the city. They brought a level of credibility to the station that a typical DJ could never do. When DeRogatis and Wyman talked at length about artists like the Boredoms and Steve Albini, it was clear that there wasn't anything particularly edgy about Q101 playing "Head like a Hole" three times daily.

And then there was *The Local Music Showcase* (Later *Local 101*). *The Local Music Showcase* launched in 1993, and two brief cancellations aside, it lasted until Q101's end in 2011. The disc jockey-programmed, Chicago music-focused show had only three hosts through the years: Carla Leonardo, myself, and Chris Payne.

Just as *The Local Music Showcase* was getting off the ground, the Chicago music scene was absolutely *exploding*. Nineteen ninety-three was the year when *Billboard Magazine* famously and audaciously named Chicago "Rock's New Cutting Edge Capital." Since Q101 had a show that supported local music, Q101 was able to draft behind Chicago's momentum.

WXRT had a local show as well (*Local Anesthetic*, hosted by Richard Milne), but it was only 30 minutes long, and it frequently aired very non-alternative songs from the blues and jazz communities. Q101 had an opportunity to own the Chicago music scene, at least perceptually, on the air. With that in mind, Q101's *Local Music Showcase* was all opportunity and promise when Carla Leonardo started hosting it. The show got off the ground easily enough. As Carla explained to me, "I went and asked Bill (Gamble): 'this is what I'd like to do. It would give us some nice 'cred'; we'll get some advice from the community about this,' and that was basically how it started."

And Carla was the right person to bring the show to Q101's airwaves back then. To hear her tell it, "I am a frustrated A&R exec, okay. To me, music was all about taking you someplace new—to be able to say, look, I found this sound, this record, this band, this song, and I think it's really cool, and I think you're going to like it a lot—to be able to introduce people to different things. There were so many great bands in Chicago at the time. The Lupins, Catherine, Eleventh Dream Day, just so many bands that were out there and waiting to be discovered. Dolly

Varden—oh, man, the talent pool was just astonishing. It was a wonderful time to be doing a local music showcase in Chicago, let me tell you."

I started interning in Q101's Programming Department in June 1993. A major component of my internship was working with Carla on the *Local Music Showcase*. Being new to radio, that environment spoiled me totally. The show wasn't driven by a playlist; Carla was able to play any band and song she wanted, and interview any guest of her choosing. Once you work in that type of environment, traditional, heavy-rotation radio feels like a fraud.

Carla resigned from Q101 in 1995, leaving the *Local Music Showcase* job wide open—and I really mean wide open—no one at the station wanted it. I think that Robert Chase could have done a nice job with it, but he definitely wasn't interested. With no one with experience or name recognition standing in my way, I walked into Bill Gamble's office and said, "I want to host *The Local Music Showcase*."
"Have you ever been on the air?" he asked.
"Sure, I've done college radio," I replied, hoping like hell he wouldn't ask for an aircheck cassette sample. I'd done some shows on Columbia College's urban dance station WCRX, awkwardly back-announcing house music in my very non-street delivery style. Long before my *Local Music Showcase* pitch, I'd unspooled and destroyed every cassette sample of my 'CRX work, fearing that future generations might actually hear it.
Bill said, "Well, I need to hear how you sound. Can you get me tape?"
"I can do that. I'll get right on that." He actually called my bluff.

My strategy was to put together a scoped (music edited out) mocked-up aircheck of my vision for *The Local Music Showcase*. To make it interesting, I went out and recorded interviews with a few bands at local venues. I talked to Wesley Willis—rest his soul—at Lounge Ax, Loud Lucy at Fireside Bowl, and the Nubile Thangs at Elbo Room (I think). After spending countless after-hours in the station's rundown production studio with a stack of reel tapes, I spliced together the interview highlights and voiced my demo. I gave the tape to Bill less than one week after he told me to give him a tape.

I got the gig shortly thereafter, but I don't think it had anything to do with the quality of the tape or my vision. It really had more to do with the fact that no one at the station wanted to do the show.

Beyond that, I knew that the show wasn't exactly a big priority for the station. It was buried in the Sunday night line-up, rolling out of *Sound Opinions*.

"On a radio station like Q101, local music is really important," Gamble told me. "I'm enough of an old school PD that it's, yeah, I know that's important and

I get why we do it, but … I guess to use a grocery store analogy, it's a loss leader. I know I'm not going to get good ratings from this show, and probably the same for *Sound Opinions*; but they're important because they bring the kind of people to the radio station that are the trendsetters, the influencers, that help keep the cool edge of the radio station. (Those listeners) weren't the "I want to hear 'Hand in My Pocket' and 'Everything Zen'" (listeners). A different group of people listened to those shows. And it helped connect us to local bands. A lot of those local bands became big, you know, due to the radio station; they helped us and we helped them. So, yeah, it was good business. Again, from a ratings point of view? Hard to say, but, you know, I think Q101 was a big enough radio station at that time that we were able to do those kinds of things. We didn't have to micromanage, 'Oh, well, if we go down to a four (share) this hour, it's going to…' It wasn't like that. We were so big that we could afford to do those kinds of things."

The Local Music Showcase was briefly cancelled at two different points in the 1990s. The news of the first cancellation couldn't have had worse timing. I made a habit of pre-taping band performances for the show during normal weekday business hours, when I had the benefit of a sound engineer on site (the talented and garrulous Vic Drescher).

Bill Gamble called me into his office and politely and professionally explained that the show's ratings were dragging the weekend ratings down, and that the station was pulling it off the schedule immediately. It had nothing to do with me, I was told; it was just business. The whole conversation took five, maybe 10, minutes.

On that particular morning, I'd booked a noisy little band called Loraxx to come in for a recording session. When I walked out of Bill's office, I checked my watch. The band was supposed to arrive in minutes. I had to tell them to turn around and leave. Suddenly, I heard my name paged on the intercom; the band was waiting for me in the loading dock.

I'm sure the band members had ditched out of work to come down to the studio. They were probably looking forward to the added exposure, too. After shaking their hands, I said awkwardly, "Um, the show's just been cancelled. We can't do the recording."

After that morning, I fell out of contact with the band. They probably hated me (consider the messenger killed), but I was also too embarrassed by the situation to go chasing after them once the show was revived only a few months later.

In 1998, *The Local Music Showcase* was renamed *Local 101,* to coincide with the release of Q101's CD compilation of the same name.

The fact that my involvement with the show continues to define my on-air work is fine by me; I had more fun hosting the show than anything else I've done in my FM broadcasting career. I interviewed Liz Phair, Veruca Salt, Smashing Pumpkins, Eleventh Dream Day, the Sea and Cake, Wilco, Die Warzau, and My Life with the Thrill Kill Kult. My show featured in-studio performances from artists like the Jesus Lizard, Local H, Wesley Willis, Smoking Popes, Red Red Meat, Triplefastaction, Verbow, Diane Izzo, OK Go, and Loud Lucy. Ratings be damned, the show was a helluva lot of fun.

I left Q101 in December 2000, to take a job at WXRT (a ten-month experience which is worthy of a book all its own). On my way out the door, I handed the reins over to Chris Payne, who had previously hosted the *Chicago Rocks* program on Rock 103.5. Payne officially took over in 2001 and stayed in the chair until 2011.

MOUTH FOR WAR: ROCK 103.5

CHAPTER SIX

By the mid-90s, "alternative music" was a buzz term that turned into a kitchen sink, all-inclusive, pop culture category that encompassed pretty much anything that was not country. The alternative radio format didn't differentiate between underground artists like Sebadoh and slick popsters like Alanis Morissette. The station's philosophy was, "If we say a song's alternative, it's alternative." With that much latitude, alternative was a hard format to screw up.

Keeping Q101 on its toes was the fiercest competitor the station would encounter during its time on the dial, Rock 103.5 (WRCX).

"The Rock" was anchored by morning show host, Erich "Mancow" Muller, a larger-than-life, rapidly rising, personality whose "in your face" approach was the opposite of Q101's more benign (and sometimes vanilla) approach.

At Rock 103.5's helm was a tenacious, personality-focused, Program Director who treated The Art of War *as his playbook: Dave Richards.*

DAVE RICHARDS *Program Director, Rock 103.5*
When I got to the Rock, it was a brilliant company (Evergreen), run by 'creatives' who believed in the product and who said, "Create a radio station that is all about attitude and fun, and has a whole lot of swagger." And we did.

FREAK *Air Personality, Rock 103.5*
We were a gang. It was us against the world. That's the only way to put it. It didn't matter who you were. If you were LITE FM or NPR, you were the enemy. It wasn't just Q101 when we were at Rock 103.5; it was everybody.

DAVE RICHARDS

Bottom line: radio people don't go in it necessarily just to have fun; they go in it to win. We are a competitive medium that wants to be on top, and number one. So, our goal was to beat Q101 in the respective demos any way we could: first by talent, and then hopefully, by music. ... These stations were neck and neck. One quarter, Q101 would beat the Rock, another quarter.... The Rock was always a male-leaning radio station; so, we would win the male demos. We were also a little older, so the new music was certainly going to have a bigger impact on a younger demo. We couldn't compete with the teens that Q101 had, and the young 20s, but we certainly could on an older audience. It was an awesome battle.

BILL GAMBLE, *Program Director*

It was junior high with transmitters. And it was ... Rock 103.5 would announce this, and then we'd announce that, then Rock 103.5 would say, "We're gonna do something with some band." Then, we'd call up and say, "Well, that's fine, but if you do that, you'll never be played anywhere in America again!" It was just that we were both doing things to try to protect our brand, and, Mancow was Mancow; ... we destroyed them after 10 o'clock, but that's pretty historic for most Mancow radio stations. (Mancow's show) was big in the morning because it was a talk show. And it's tough to build a music image when you have a talk morning show like that.

DAVE RICHARDS

At the time, alternative was blowing up. I moved from Seattle, the capital of music of the time, the Haight-Asbury of the early '90s, to Chicago, and Q101 was a radio station that was all wrapped up in everything Seattle and every band that was blowing up alternative. We had to create a station based on rock, and rock at the time was a little bit of this, a little bit of that -- a little bit of current rock, a little bit of alternative, a little bit of classic rock. We weren't about to beat Q101 on a music to music battle; we had to beat Q101 on an attitude and fun and entertainment level, which is why we were very talent heavy—maybe musically disjointed a little bit—but very talent heavy, and very talent focused.

CHRIS "PAYNE" MILLER, *Air Personality, Rock 103.5*

I remember sitting in meetings at Rock 103.5 with Dave Richards, Ned Spindle, and a few others and we would all go out, smoke a bag of weed—not a bag, but we'd all party, sit down at a restaurant and brainstorm on how we could beat Q101. All guerilla radio: how we could just be in their face, how we could steal every show, how we could get the "presents" (station sponsorship) on a show, and how we could show up at their events and crash their parties. It was just a battle. It was really a full-blown battle. And the battle was on the air, too; I mean, ripping on the other jocks. It was just in their face.

I think, with the exception of me and James VanOsdol, the *Local 101* thing, and *Chicago Rocks*—we didn't have that competition.

DAVE RICHARDS

At a certain point, five years in, we came up with sort of a matrix of how maybe to put a real big nail in the coffin of Q101, and that was to do a tight focus on men in their mid-20s, because that's where our talent was and that's where the heart of the radio station was. I think we were focusing on a 25-34-year-old male, and it was magic. It worked.

CHRIS "PAYNE" MILLER

I've since become friends with Dave Richards and have an incredible amount of respect for him. Working with the guy at Rock 103.5 was pretty amazing. He was a really talented guy and he had a lot of respect for his staff, the talent.

SLUDGE *Air Personality, Rock 103.5*

I would say it was almost unlike anything since. I'm talking about true war on the streets. If there was a promotion, we always felt, with that gang mentality, that if they're over there, we're going to show up over there, and see what happens. It never got physical; it was always fun, verbal terrorism, if you want to call it that. Rock 103.5 had stickers, but they didn't even say Rock 103.5; they just said Q101 Sucks.

FREAK

On my garbage can to this day, there's a Queer 101 sticker and a Queer 101 Sucks sticker. I remember we had those made, and used to hand those out. It was a good old-fashioned war: all the little digs, trying to get the different shows. Trying to get the different shows was a big deal. It was a big deal to use the words "presents" or "welcomes." You wouldn't believe how much blood was shed over those different words. It was just goofy. To me, I don't think the average listener knew the difference, but to us in the halls …

SLUDGE
It felt like an everyday adrenaline rush; like, how can we fuck with somebody today?

ZOLTAR, *Air personality*
They were obsessed with us, which was cool. I liked that.

SLUDGE
It was that Chicago mentality, like, "Hey, we're going to go over there. We're going to handle our business and take care of things." It was a blast. The passion to win on both sides was awesome. Back in the day, that created better radio for the listener. I think, at the end of the day, it hopefully benefitted them. My gosh, especially the battles … and having Mancow as your general was a little comforting, I guess. Rock 103.5 was a pure attitude station. Q101 was a powerhouse of what it was at the time, one of the best alternative stations in the country.

MIKE ENGLEBRECHT *Promotions Coordinator*
I remember as an intern going out to events and the Rock 103.5 crew was there. I vividly remember being at the U2 shows as an intern, and at the end of the show, handing out stickers. Ten feet away from me, it's funny, I remember loudmouth Jim Lynam handing out stickers, barking at me, and me not saying much.

KEITH SGARIGLIA, *Promotions Assistant*
Their promotions people used to follow us around and take pictures of what we were doing.

CHRISTIAN "CAP" PEDERSEN, *Promotions Assistant*
Walking into the station on my first day, I could instantly sense everyone in the Q101 Promotions Department lived and breathed the station: the music, the culture, and the lifestyle. I quickly learned interns and promotional assistants were the face of Q101 on the street, at events, and at concerts. There was no room in the Q101 Promotions Department for people who didn't believe in and bleed

Q101. We always felt as though we were the underdog, constantly going up against the aggressive Rock 103.5 and its listeners. Every time an artist or group that was played on both stations came into town, the interns and the promotional assistants would trek out to Alpine Valley, The New World Music Theatre, or whatever the venue, play music from the van on the way in, and distribute Q101 static stickers on the way out. That was always the dicey part of the event, when we had to pass out stickers. We would get either those who loved Q101 and requested multiple stickers, or those who were P1 listeners, and listeners only of Rock 103.5. Those were the listeners who would always find delight in shouting, "Hey, it's Queer 101! Go fuck yourself." We would always have to keep our cool, and with tongue in cheek, say, "Thank you so much. Have a great night."

WOKE UP THIS MORNING: Q101'S MOST UNSTABLE AIRSHIFT

CHAPTER SEVEN

Morning Drive was radio's "money shift" for decades running, and especially so during the 1990s. To Q101 management's chagrin, the station's morning drive slot never quite clicked with advertisers or the audience for the bulk of the 1990s—not that efforts weren't frequently made to turn things around.

ROBERT MURPHY, *Air personality*

I always listened whenever they changed the morning shows—that was, what? Nineteen times? I tuned in each week to see who the new morning show was. That interested me.

CHUCK HILLIER, *General Manager*

It was a whirlwind, it was tough, and all it really said was that people were really there for the music, but that started at 10 o'clock. Nothing we were doing really had an impact on the mornings. That's entirely typical with music-driven radio stations.

REY MENA, *Marketing Director*

There was no consistency. In my opinion, there was a lot of trial and effort, and pulling the plug early on stuff.

ERIC "SHARK" OLSON, *Air personality*

I remember management seemed to be always looking for a morning show, and seemingly, testing out different people. When the show wouldn't quickly take off, the station would immediately move on to a different setup. It seemed the Q was very impatient when it came to organically growing a local morning show.

CHUCK HILLIER

I look back at that and it was beyond a revolving door, for crying out loud … I didn't like trifling with people's lives. I'm not sure that there was anybody that was not giving it his or her best. Everybody was sincere; everybody wanted to win; everybody had their individual strengths, but they were put into a very untenable position, honestly. They were all competent, great, good folks, and yet, there wasn't anything in subsequent ratings that would indicate that we had found the magic pill.

BOBBY SKAFISH, *Air personality*

After a show one day, I got called into the office, and Bill said, "We're making a change. We're going in a different direction,"—just vague terms like that. He offered me to stay on part-time, which I wasn't about to do. After he was through with me, which didn't take long, he goes, "Maybe you want to stop in and see Chas," which was a nickname for Chuck Hillier. I declined on that. I took five minutes to clean out my personal effects, and I split, and that was it.

BROOKE HUNTER, *Air personality*

At that point, Bill really didn't allow a morning show much more than six months. I don't think he gave Skafish more than 6-8 months.

BOBBY SKAFISH

It was almost like the 'self-helpy' books from that point—I might've had a little more swagger, a little more fake-it-till-you-make-it, kind of… I think, maybe, I was emotionally at a little bit more vulnerable point of my life, with the Loop having had a format change. After having been there for ten years, all of a sudden I'm doing overnights on Q101. I think that maybe, it was not the strongest period of my life, emotionally. I might have had a little more presence and command, and maybe a little more swagger, but that was where I was at right then. So, in a way, I went, and said "No regrets."

ROBERT CHASE, *Air personality*

It was a week that I'll never forget, because I think it was on Thursday I signed off on my first house—bought my first home. Anybody who's done that knows what a deep breath you have to take, to take on that responsibility. Friday, Bill tells me, "I want you to do the morning show." I told him, "I don't want to; I've never done a morning show." He said, "That's exactly what I want to hear. I want somebody who doesn't know what they're doing to do it." He just thought that not having the trappings or experience of doing a morning show might free me up to do … something … but I honestly liked doing nights; I was really happy.

I didn't want to do it, but there was no way out. It was what he wanted. And then, the next day, I got married. So, in three days, it was just a lot.

Any experience I had had filling in on mornings on the way in my career, you did everything yourself; maybe, you had a newsperson. I had to learn the ropes and the rhythm of throwing it to the traffic, throwing it to Whip (traffic), going to Heidi Hess (news) … all the different moving parts of a morning show. Every quarter hour is kind of mapped out as to what's going to be taking place—either someone's going to be on the phone for an interview, someone's going to be walking through the door for an interview. The preparation that goes into a morning show is just something that I had to get up to speed really freaking quick. I tried, and I think we struck a nerve. There were times, and I remember Bill calling me after one show saying, "You know that was really just too small a market today." And he was right. But sometimes people were coming through the door who I didn't even know were going to be coming through the door, or I forgot they were going to be coming through the door. So, it was imperfect, but I think it was endearing on some level.

BILL GAMBLE, *Program Director*

In retrospect, probably, having Robert do mornings – it was one of those decisions I wish I would've just left alone. If given time, you know what? Look at Lin (Brehmer) at XRT. That was a combination of talent and time. And ultimately, in radio, it's a thing -- today we have even less, because of almost daily ratings. But back then, we were … the good thing is that we reacted quickly to musical trends and bands and things. The bad part about that is, you got caught up into that change, and you wanted the instant, "Well, wait, we're #1 mid-days, we're #1 afternoon drive, we're #1 at night. Hey, we're not #1 in the morning. Let's try something." And, looking back through the years, I think Robert Chase, as a morning person, with the right stuff, I mean, he could've continued.

BILL LEFF, *Air Personality*

That was a time where there was a revolving door on the morning show, and if you were driving past the Merchandise Mart, they would let you do the show for a few weeks.

I think that was how Lance and Stoley got the job; I think they were delivering a pizza and (Q101 management) said, "If you guys wanna be on the air, feel free."

'Lance and Stoley' was one of the more polarizing shows Q101 put on the air in the 1990s. The titular talkers went on the air with no experience whatsoever, unless you count the handful of times their band, The Lupins, was interviewed on the Local Music Showcase with Carla Leonardo.

STOLEY, *Air Personality*

Lance had approached Carla because of the *Local Music Showcase*, and had gotten us on to that. And then I remember we got to go back on the *Local Music Showcase* when we got our record deal, because we were interesting; so, we got a second visit. From there, I think Bill heard us. I think Bill was like, "Wow, those guys are funny." Carla, she was the one. If she hadn't supported local Chicago music, we never would've been on that showcase, and then we never would've got our weeknight show, and then, of course, the subsequent ruling of the airwaves in the morning.

LANCE TAWZER, *Air Personality*

Carla came to our gigs; Carla listened to our records. Carla cared about the local music, and we love Carla. We worked on those *Random Acts* CDs with Carla. She was real to us. We kind of felt like, she was in it for the right reasons. We felt that she was somebody who really was out there for the local music.

CARLA LEONARDO, *Air personality*

I thought they were a great band. I'm sorry, but I still think that "Transparent Faye" should have been a big hit.

LANCE TAWZER

We were trying to be funny, and trying to be entertaining, and we did a couple of in-studio things. That was fun, and the way the story goes is that we had an opening act slot with my old band, Material Issue, at one of those block parties. It was Material Issue one night, and Counting Crows on another night. We opened up for Material Issue, and Bill Gamble was there, and we knew he ran a radio station, so we chummed up and tried to be funny, and tried to make him like us, in some way, shape, or form, because we thought it might be beneficial to our careers some day.

JON REENS, *Research/Promotions*

I think that the reason they were brought on was that they weren't radio people. As soon as you start managing them, they become radio people; they lose their edge.

LANCE TAWZER

We got a call from Bill saying, "Would you be interested in coming in and doing a show on Friday nights from 10-midnight?" I can't remember how much money he offered, but at the time, we were starving artists, and whatever it was, it was a lot of money to us. We said, "Sure, 10-midnight. You're going to pay us that amount of money? What are the rules? He didn't really lay down a particular plan; he wanted us to be pretty freeform. We had Carla there to keep us in line, more or less.

BILL GAMBLE

When you do research, the audience always comes back and tells you, "I don't want to hear songs repeated. I want to hear all this different kind of music." And, if you were ever to put a radio station on like that, you wouldn't have any listeners. Because ultimately, people want to hear new good music, and they want to hear songs they know and like. And I think it's a lot (like that) with the DJs. Lance and Stoley were really funny, and they were really great at 10 o'clock on a Friday night when you're out partying or just wacked out at your apartment somewhere.

CARLA LEONARDO

At the time I was like, "What the hell?" In retrospect, things were changing so rapidly in radio that it was probably Bill Gamble who was pushing the envelope, and he did it in a lot of different ways. It was not a way of doing radio that I was used to, or comfortable with. When your boss tells you to lose some brain cells over the weekend, some people like that; others, like me, don't. At that point, I was becoming obsolete in the format.

LANCE TAWZER

It was pretty loose and pretty crazy, but one night a week, we didn't think much of it.

BROOKE HUNTER, *Air personality*

I was still doing my night show, and they could come in for the last hour of my night show. I wasn't necessarily part of the show at first; they had me there as the person who could push the buttons. Then, as we got to know each other a little bit better, they would start bringing me in and we would talk, and that was kind of how the Lance and Stoley morning show came to be.

LANCE TAWZER

Ten to midnight, nobody was there. You ran around; you'd try to steal things out of the prize closet; you had the run of the place. I was always hoping that Mary Shuminas would have her door open so we could steal CDs, because she always had tons and tons of stuff.

STOLEY

Nobody was, like, "Oh, we met those Lupin radio guys, and they were assholes just like we thought they were." We always ended up friends. When Veruca Salt was on our show, we had a great time; we all ended up being great friends. Urge Overkill—to this day, people from the Lupins talk to people from Urge Overkill.

BROOKE HUNTER

We had Liz Phair up for a long time at night with Lance and Stoley and me. That was always fun; it was more like hanging out with a neighborhood friend, considering she was from Chicago. Lance and Stoley would go up there and get stoned in the newsroom, and on Monday morning, Bill would walk in with his suit and his briefcase going back to work. He would walk in to say hello to Lance and

Stoley, I would look over on the news computer, and there'd be a big joint sitting there. I'm like, "Oh, come on, this isn't good." Bill walked by it; maybe he saw it, maybe he didn't.

BILL GAMBLE

They had been doing the late night, 10 o'clock to 2 a.m. Friday night—let's put a band on and be bizarre, and Brooke was there to babysit. And then they got Liz Phair in, and it was pretty entertaining radio—not the kind of stuff you hear everywhere. And, there was a real chemistry. The good thing about being successful is that … it encourages you to take risks. The bad thing about being successful is it encourages you to take risks that sometimes you don't need to take. And, Lance and Stoley were perfect—it was like Zoltar: great DJ, perfect DJ, the perfect thing for a radio station like Q101. Because when Q101 became really big, and had mass appeal, and we were one of the first FM radio stations to 'cume' a million listeners, the cool thing became, from that leading edge of the curve, to bash Q101. I mean, we all remember being at the venues where it was, "Eff you! Eff you!" … it was cool to say, "I didn't like Q101." But it was also … we had to give them something to say that they really liked. And that's why Zoltar was on the radio. "Yeah, Q101 sucks, but *The Industrial Zone*, no, that's cool.'"

STOLEY

Zoltar claimed to be from another planet … I remember one morning, Lance was talking to Zoltar and he said something about "your people," or "what's it like with your people," and Zoltar was, like, "What do you mean by that?" because he's a black man. Lance was like, "You know, on your planet," and Zoltar was, like, "Oh that's right; I'm from another planet."

LANCE TAWZER

We used to be up on the 17th floor, and we'd have to buzz (people) in. I would have Jim Ellison and Liz Phair and guys that were in bands that we were buddies with, and we'd have them on the radio and we'd be pretty loose, and then we'd run into Zoltar, who was, like, the coolest dude on earth. He would light incense, and he'd have all these women calling him. We thought, "This guy really has the gig."

STOLEY

Those were the good days. We didn't know what we were doing, and we were just making stuff up and having our friends call in; kids were calling up. It was an exploratory time. It was fun because there was nothing at stake. I didn't feel as though I had anything to lose. We were just so lucky. It was weird.

LANCE TAWZER

A series of lunches happened. Bill and Chuck Hillier took us out to lunch one day and said, "We like what you're doing, and we want you to do it five days a week."

BROOKE HUNTER

At that point, we could kind of read Bill. Anything Bill started, because we had no morning show at the time, was always a possibility for the morning show. I remember saying to Lance and Stoley, at one point, when we were doing the night show, I had a feeling they were going to move us to mornings—and ultimately he did. He wanted all of us to do mornings. Lance and I were in; we were good. Stoley did not want to do the morning show; he was adamantly against it. Right around that time, we had Jamboree coming up …. Bill took me aside before Jamboree and said, "Brooke, I don't know what you can do but I need you to help me convince Stoley to do this morning show."

STOLEY

Money: I don't think it's a misconception anymore, but I think, in those days, it was probably more of a revelation that when you get a record deal, you don't get a million dollars handed to you, and your record doesn't come out the day after they sign you and all the stars align. While we had a record deal, a lot of the guys in the band were still working jobs, still trying to make ends meet, and trying to figure out how we could go on the road and still support ourselves. It was a lot of money, and it was cool. Being courted was cool. When they took us out to lunch and we got free margaritas and burritos, it was all very cool. Chuck Hillier and Bill Gamble did that. I think so much of this. Lance was so much more privy and so much more involved in everything, and it was kind of like, "Stoley, they're taking us out to lunch. I think they're going to offer us the morning show."

LANCE TAWZER

We had another lunch with Chuck and Bill. It was Mancow—Mancow was killing everybody. Mancow had his own shtick, and Bill was trying to figure out a way to beat him. Apparently, he was thinking that we could be somebody who would attract his audience away from Mancow, that we could kind of be an alternative to Mancow. He was looking for the Wayne's World or Beavis and Butthead partnership. He took Stoley and I to lunch, and I remember him saying, "We've spoken to your record company and management, and they're really excited about this. They think it's a good bio piece for your band." …. We walked away from lunch thinking, "Oh, god, what do you think?" We knew how the band was going to react—they were going to think, "Don't do it. Nobody takes morning shows seriously."

STOLEY

I think I was concerned about losing credibility. I think there was already some backlash in the music scene, be it jealousy, or be it that people truly did find us annoying on the radio. There were other bands and cool people in the city that I think saw us a bit of a joke, an invention of a radio station. It sucked, because I had really worked hard on my music … the whole fight within me, in the whole span, was that I wanted the music to be taken seriously, even though I wanted people to love me as a hilarious goofball. I think that turmoil was obviously present, even before the morning show started. If I was resistant, that was absolutely why.

LANCE TAWZER

We felt terrible, because it meant Robert would have to go back to evenings, or to whatever gig he was doing before he went to mornings.

BROOKE HUNTER

It was Stoley just being lazy and not wanting to get up at 3 o'clock in the morning. Bill found the money and then got Lance and I to kind of say, "Stoley this would be a really good idea—you would be getting exposure for yourself, you would be getting exposure for the Lupins, and it would just be something that would be worth your while to at least try out for a while. And once again, (the morning show) lasted 6 months.

BILL GAMBLE

It was like any negotiation, "No, well that's fine. I don't want him to do it." "No, no! We want to do it!" And eventually, he decided he wanted to do it. And it was—it was a spectacular failure. You know, if you're gonna fail, fail really big. And with Lance and Stoley, it was a one-in-, I don't know; -pick a number, shot to happen. But it was I feel blessed I was able to work for a company that let us try stuff like that.

LANCE TAWZER

We were going to make the union minimum, which I think was like 40 grand a year, which was a lot of money for us. Initially the band was, like, "I don't know if you really want to do it." But we were, like, "We've got to do it. It's regular money." We had already signed our deal with RCA records. In the midst of making our record, we were not making any money, so we said "Okay," and that's really, where we developed what the show could be. We just kind of decided immediately, "Let's not pretend that we're DJs. Let's not try to come up with a 'thing.'"

BROOKE HUNTER

Oh, they were a train wreck, but I think it was a good concept. If Bill had given it a little more time and put more into it himself, it could have actually become something. I don't think it could have ever been (like) Kevin and Bean (on KROQ), but I think it could have become something. But he didn't want to give it the time.

LANCE TAWZER

Brooke was essentially there to reel us in, and she did her best. We did our best to challenge her in every way we possibly could.

STOLEY

Brooke was like my homey. She used to drive me home in her little red sports car. She took me to buy a TV once, because I was making radio money and I was going to buy a big TV. Brooke knew how to buy stuff.

ROBERT CHASE

Lance and Stoley had charmed a lot of people, and there was certainly something about them I couldn't even deny myself. And they were becoming friends of mine as well, so there were no hard feelings about that. And I think when Bill said he was going to make some changes, and I asked if I could still stay on part-time, I think that was kind of an odd moment for him that someone was really being let go but saying, "Can I stay on in a lesser role?" Maybe my ego was just happy to be part of it.

STOLEY

We had "Cop Stories," and we had "Stoley is a Lying Bastard." Those games. I guess the high point of the games was when we dismissed the games and just said, "What will you trade us for prizes?" We had Game Boys. We were, like, "We'll have no competition. If you want to trade us something interesting, we will trade you for this swag that we have." Somebody traded me a human brain.

The woman that gave it to me claims that she had limbs and stuff; she had worked in a morgue. She had other body parts, should I be interested.

I think this is a good analogy of my experience at the radio station. It was kind of like, "Hey, I'll give you a radio show!" And you're, like, "Yeah, that sounds great."

"I'll give you a human brain." "Sounds pretty cool." And then you see a human brain on your desk, and you're, like, "Holy shit, that's a fucking brain."

LANCE TAWZER

They wanted us to be ourselves. They wanted us to be musicians. They wanted us not to be DJs. It was a lark. It was a crazy idea of Bill Gamble, to try to combat Mancow. He was killing them. Robert Chase hadn't worked out. Whoever was before Robert hadn't worked out.

STOLEY

Those shows, the whole thing, everything from the Friday nights on up to the morning, for me, were fairly similar. I remember going to the Merchandise Mart with nobody there, ever. It was dark and weird, and Lance and I played Super Nintendo.

LANCE TAWZER

What (Bill) wanted was for us to come across as petulant, non-radio guys that he was hoping people could relate to.

STOLEY

Lance, I think, was probably more, like, "Wow, we get to be DJs. I'm going to read the sports." And I was more, like, "You're trying to sound like a DJ," thinking the whole thing was ridiculous.

NOT FOR YOU: Q101 VS. PEARL JAM

CHAPTER EIGHT

Q101 was slow to embrace Pearl Jam, who within four years of their debut became one of the most important bands of the 1990s. By the time Pearl Jam was scheduled to play Soldier Field in the summer of 1995, Q101 was playing the band's songs on a 60-75 minute rotation.

Perceptual ownership of artists is something radio stations aggressively try to attain, especially when faced with a strong competitor in the market (in Q101's case, Rock 103.5). Shortly before the Soldier Field concert, Q101 launched a billboard campaign that appropriated a Pearl Jam lyric/song title from the then-current Vitalogy album: "Not for You."

BILL GAMBLE, *Program Director*
Billboards came up, and it was like, "Oh, let's use 'Not for You.'" I think it was a lot less litigious then, than it is now. We put 'em up and it was great. There was great buzz on the 'boards.

ALAN SIMKOWSKI, *Director of Market Development*
Rey Mena was Marketing Director at the time, and his team thought that using that lyric would be a really good idea.

REY MENA, *Marketing Director*
"Not for You." We had not done advertising, any advertising. It was just on buzz and word of mouth, all that stuff. Finally, we had the opportunity to do an

advertising campaign. We opted to try to do something that spoke to the audience, while sending a completely different message to the general public.

MARY SHUMINAS, *Music Director*

I guess it was kind of an anti-advertising campaign.

REY MENA

It played on two levels because if you listened to Pearl Jam, you knew it was a song, but if you didn't, you had a station basically saying it was not for them. After we put the billboards up, there was a major Chicago advertiser who actually called the station yelling at us, basically saying, "How can you advertise the station's not for them? You're insulting the listeners." He didn't understand it was a Pearl Jam song.

MARY SHUMINAS

It was us trying to position Q101 as an elite station.

BILL GAMBLE

Soldier Field, Pearl Jam comes to town.

KEITH SGARIGLIA, *Promotions Assistant*

It was one of the best nights I had because I was in school still. I came down on the train. Brian the Whipping Boy and I were up at dawn, driving out with a pair of tickets, the morning of the show — out to some strip mall. There were probably 300 people there within about 5 minutes of us calling in, and we had boxes of "This is 'Not for You'" t-shirts. We gave all those out, gave away the two tickets, and drove back to the city; then I remember getting a tape recorder so we could record the shortwave radio (broadcast).

JON REENS, *Research/Promotions*

Pearl Jam was the first show I was remotely involved with. I was working in research. I was there five days a week, from 8 am until 5 pm, in this room, no windows, barely any air conditioning, getting barked at. Steve Levy frantically came into the room and said, "I need somebody to go down to Soldier Field. I need somebody to record this; they're broadcasting it!"

I piled in the van and went down to Soldier Field. I had never done a promotion for the station, never been involved in anything like that. I merely did call-out research for four hours a day. I sat at the van with a transistor radio and tape deck and a bunch of cassette tapes, 45 minutes each side. I would record it, and have to make sure that I found a break in the music to flip the tape so I didn't miss the songs.

I just sat there and flipped tapes. I had to bring (the tapes) in that night, I had to go back to the station. Bill Gamble had me sit in there for the first part of listening with him and Rey to make sure that it was right and good.

BRIAN "THE WHIPPING BOY" PARUCH, *Air personality*
We ripped off one of (Eddie Vedder's) lines and put it all over Chicago. When Pearl Jam headlined Soldier Field in 1995, Q101 flew a plane over the stadium with the billboard image trailing behind it.

MARY SHUMINAS
We strategically placed (the plane) so the band would see it.

REY MENA
That was when Eddie Vedder was at his anti-establishment peak.

EDDIE VEDDER (onstage at Soldier Field, July 11, 1995)
"Today we were all watching Otis Rush; it was just a beautiful day. I was just into it, and then there was all these flying bugs with banners with radio stations' lame-ass messages on the back, fuckin' up the view. I'm against guns and military spending, but a big, big piece of something would've been great, right about then.

"One of the radio stations said they had a two-hour Pearl Jam show afterwards. Well, if you listen to 89.1, we got a radio station out of our van in the back (and you can) get the real shit. I think we're broadcasting right now, this whole show, out of the van, to Chicago. All of the people who couldn't get in, or thought it was going to suck because it was such a big place, well fuck 'em. I understand; I know where they're coming from, though. I guess there's a big billboard that's every-fucking-where that says, '"Somethin', somethin', this is not for you."' The joke's on them, because it's not for them."

BILL GAMBLE

Eddie, basically says, "F the radio station."

BRIAN "THE WHIPPING BOY" PARUCH

(Vedder) got mad at our plane flying by … then you get these 50,000 sheep cheering every bit of that "Yeah! No military spending! Yeah! We hate Q101, too!" He actually hated our radio station, and then we put up billboards that said, "This is not for you." We ripped off one of his lines from one of his songs, and put it all over Chicago.

BILL GAMBLE

I'm thinking, "60,000 vs. 1; I'm going to lose this thing." It was like, "Whoa, here's, at the time, arguably one of the biggest bands in the world, calling out your radio station. Pretty cool."

REY MENA

We actually thought it was cool that he was 'dissing' us.

STEVE FISHER, *Air personality*

I thought, "Okay, this is good and this is not good." This was basically saying—Eddie, being Eddie, being the rebel that he is, was basically saying, "We're not a trademark. We're not sell-outs. We battled Ticketmaster. We're certainly not going to let a commercial radio station use our song title to brand themselves." I think he was basically saying, "F you" to mainstream media, and that made him cooler with all of his followers, so to speak.

VANCE KORETOS, *Research*

I was at the Pearl Jam Solder Field show and remember Eddie's rant about the "Q101 This is Not for You" billboards, stickers, and the plane flying with the banner. I was on Pearl Jam's side of that and thought it was an odd choice for a slogan campaign, but it worked, I guess, because we all remember it.

KEITH SGARIGLIA

What do they say? All promotion is good promotion, so I think that wasn't bad. Of course, everyone booed us, I think, when he did that. But there was some uneasiness—you don't want the lead singer of the most popular band in the world 'dissing' on your station.

REY MENA

All of a sudden, an interesting thing happened that was a telltale sign tied to that event, and that was the rise of Mancow. We were all excited about the show and about what we were going to do after the show, poking a finger in the eye of our competitor. We were sitting there with all of our team, and all of a sudden, we hear all this rumbling, and people looking to the right. We thought it was a fight or something, but it wasn't. It was actually Mancow walking in with Turd … the whole deal. That was the first time it was a palpable sentiment that this guy was going to be big.

WHERE IT'S AT: JAMBOREE AND CRITICAL MASS IN THE MID-90s

CHAPTER NINE

After the success of Q101's first major radio show (Twisted Christmas in 1994), the station added another event to its annual calendar: Jamboree, an outdoor event intended to launch the Chicago summer concert season. For its first multi-stage lineup, Q101 enlisted a broad range of talent that spoke to the inclusive nature of the alternative format in 1995. Appearing on the bill (in alphabetical order) were Bush, Collective Soul, Sheryl Crow, Duran Duran, Faith No More, The Flaming Lips, KMFDM, The Lupins, Phunk Junkeez, Sponge, and the Stone Roses.

ROBERT CHASE, *Air personality*

We decided to open the outdoor concert season in Chicago with the Jamboree, having the first concert to take place at the World. That was really fun because you're outdoors, you know, multiple stages—it was, like, "Who needs Lollapalooza? We've got our own thing now."

MARY SHUMINAS, *Music Director*

We had a brainstorming session. This was in Bill Gamble's office. It was probably (Promotion Director) Steve Levy writing things on one of those dry erase boards. We were just throwing out names, and I said, "What did they call that event in the Flintstones when the cub scouts had a pow-wow?" Someone's, like, "It was a Jamboree!" That ended up being the name we went with. I don't remember what the other considerations were.

LANCE TAWZER, *Air Personality*

Bill, always the wheeler-dealer, made it worth (The Lupins') while. We got a couple of really good gigs; we opened up for Filter and Everclear. We got a couple gigs at the Aragon Ballroom; we opened up for Bush and the Toadies. We did what we could. And of course, we played the very first Jamboree. We were the very first band on the tiny little side stage, before it started raining.

ALAN SIMKOWSKI, *Director of Market Development*

Jamboree, all those festivals, became annual events that were, like, to-be-seen and must-be-seen types of things within the industry.

REY MENA, *Marketing Director*

At one point, at one of the early Jamborees, out of 30,000 or so seats at the Tweeter Center, we had sold 20,000 t-shirts; so, we were outselling all the bands combined.

LANCE TAWZER

It was amazing for us. We didn't know what to expect. We didn't think anybody would know who we were. We were the first band going out, so we figured there would be like 12 people there, and people were going to be just coming in. That wasn't the case. Actually, it was packed from right on up to the stage, all the way to the back. What we were most excited about was, instead of dressing rooms, they gave everybody a Winnebago in the back parking lot, behind all the stages. And we could sell merchandise.

STOLEY, *Air Personality*

(Jamboree 95), that was great. We had a trailer; my parents had to be escorted through gated security checkpoints. There was a zillion people watching us, and they all seemed to be having fun.

LANCE TAWZER

Our biggest thrill was that our Winnebago was parked next to the Flaming Lips.

MARY SHUMINAS

I remember the feeling of the first Jamboree we did—not only did it sell out seemingly instantaneously, but the actual day of the show, we were there hours in advance, setting up obviously. As the people started pouring into the World Music Theater, the feeling of seeing thousands of people pouring into this event at the same time, I guess that was kind of a moment.

STEVE FISHER, *Air personality*

Just getting to introduce bands in front of 28,000 at the New World Music Theater slash whatever it is, Charter One Bank...place, I don't know...what they call it—that place in Tinley Park. I mean, I think that, to be in front of 27,000 people, who love the music you're playing, and are there to cheer on these artists, it just—it kind of felt like we were part of a mini-Woodstock, if you will, even though it was much less grand a scale.

BRIAN "THE WHIPPING BOY" PARUCH, *Air personality*

I remember whoever the lead singer of Sponge was (Vinnie Dombroski)—doing what I thought was a really cheesy rock star Jesus-pose at the front of the stage. He literally went out there and was, like, doing a Christ figure, cross thing, and having girls jump on him. I thought it was cheesy, but it was Sponge, so maybe he was just enjoying his moment. Maybe he knew it wouldn't be too long.

ALAN SIMKOWSKI, *Director of Market Development*

Q101 brought these bands to the masses, and took them to the next level.

ROBERT CHASE, *Air personality*

The sweet spot had to be 95-ish; it just seemed like the grunge thing had kind of played out, but was still significant with a lot of bands. Then you had that Anglo thing that was kicking in with Oasis and the Verve—big bands in '95 and '96 that we embraced and played (but) that nobody else was playing. It just seemed, like, with the Twisted shows and Jamborees, that we hit a sweet spot there.

ALAN SIMKOWSKI

We started making a ton of money as a radio station on those things, and then we really started getting the bands involved with everything we were doing.

KEITH SGARIGLIA, *Promotions Assistant*

That was the height of its popularity, that summer of '95. Just everyone listened to Q101. That led to the excitement. Everyone there was excited about working for the top station in the city, and everyone was into having a good time. They enjoyed each other's company, which I seem to remember the most.

SAMANTHA JAMES, *Air personality*

It just seemed like there was this summer where everything was Q101. Everybody was talking about the next Q101 event, or the next big concert.

BRIAN "THE WHIPPING BOY" PARUCH

Between '95 and '97, I thought we sounded pretty good.

RYAN MANNO, *Air personality*

Q101 was dangerous back in the day. It was alive and you knew it and you wanted to listen, because you thought you'd miss something (if you didn't).

ALAN SIMKOWSKI

It was just exploding, the music. We mainstreamed all that music.

HEIDI HESS, *Air personality*

I really think it (was) such an era ... it meant something. The music that Q101 played for most of the time was something that still matters.

BILL GAMBLE, *Program Director*

You know what it was—this sounds really ridiculous, but—we just listened. I had a great Music Director; I had lots of people, people like (James VanOsdol) in the music department. And, we'd just listen. We'd listen to stuff, and, at times, we'd argue about stuff. And, you know, we, at least in my case, liked any format where I didn't really know what I was doing when I started out,; it was really good, because I didn't bring any preconceived ideas. It was, like, "Oh, well, you have to play the new Cure song." "Well, no we don't, because The Cure sucks now and no one cares about them." It was, like, when you don't know what you don't know, you sometimes do a better job. And because all of what we were doing was brand new, I mean, I remember hearing "Bulls on Parade," and we just instantly put it in powers (heavy rotation). I mean, we were playing that song 70 times a week! Now, granted, we were in Chicago, but I mean, even today, that's a pretty … that's a pretty loud record. Morning drive, it didn't matter; it was "Bulls on Parade," just because we did a good job of listening to the audience, staying in touch with stuff on the street, and then not bringing preconceived notions into what we should play. I mean, Alanis Morissette. I mean, that—you look at that and go, "How did she get on an alternative radio station?" It was because the audience wanted her. And we had systems and people in place to pick up that stuff. And we really did—I've used this before, but I really believe it. I'm a customer service rep. I might have a nice office and get to go to some cool events, but really, that's what I do. Whatever they want, that's my job to give it to them. That's why I have my stack of Tom Petty CDs on my credenza. So I can go, "Oh this is what I really like." But, that's not my job. My job is to figure out what the audience likes. And so you surround yourself with really good people, that sit there and go, "You know, Bill, you don't get it." "Okay! Help me," and, that's sort of what happened. So there wasn't any real, "It's gotta be in 15 markets," or that stuff, because alternative, at that time, was top down. Top 40 was always, "Well, we're gonna start in the small markets and move up." But alternative, you know if Kevin (Weatherly, KROQ Program Director) played it in LA, or we played it in Chicago - that's all it took. It was a monster hit. There were two radio stations controlling the charts, and that was it!

CHUCK HILLIER, *General Manager*

It was a magical time.

BOBBY SKAFISH, *Air personality*

The station was successful, and there was a buzz at the station; it was an exciting time.

BROOKE HUNTER, *Air personality*

Luck and timing had more to do with it than anything. I think it was the right thing, at the right time.

STEVE FISHER

When people go, 'What was your best experience in radio?' I go, "Q101 afternoons. ' 93-'97." They're, like, "Well how come?" I'm, like, "Because it wasn't work." We were having so much fun. You know, you're going to clubs. They're paying you crazy money to go to clubs, and then, you get to go to all these concerts. And it's, like, and I can sleep in till 10? How cool is that?

BILL LEFF, *Air Personality*
Steve was a guy who lit up when he did promotional stuff. I still, to this day, call him "Join Me, Steve Fisher," because every day he was somewhere meeting people, and I think they always came away happy to know him. He's always got a smile on his face, and he's always positive.

STEVE FISHER
I've often got a lot of grief from, like, college radio students at Northwestern, who used to mock me for saying, "Join me, Steve Fisher" all the time. And, if there was any catchphrase I could have gotten rid of, I really should have, should've just said, "I'll be here. Come see me. What's up?"

JON REENS, *Promotions/Research*
We would do a weekly broadcast from Great America with Brooke, and we would make people do the stupidest things for a t-shirt, like, "Hey, go jump in the fountain." I remember where we set up; there was a thing called the Orbit, that spun and then it went up. And there was a chicken shack right next to it … the interns would drive out with all the broadcast equipment (and) Brooke would pick me up at my place because she didn't like dealing with the traffic. I would drive Brooke's car out to Great America; we would park in the back, in the employee lot, get on a golf cart, get driven to wherever we were broadcasting from, and then make people do the stupidest shit. There would be thousands upon thousands of people who would come to that theme park and stand there for four hours.

This girl said, "I want a t-shirt," and we were, like, "All right, put this ketchup and mustard in your hair." So she literally (took) ketchup and mustard bottles, and just squirted it all over her hair. And a guy came up and said, "For a Red Hot Chili Peppers CD, I will eat the ketchup and mustard out of her hair," and we said, "Okay," and he did.

STEVE FISHER
The way I looked at it, I was just like a circus emcee, master of ceremonies. (You're) just at some club, and you're throwing out t-shirts and having people

do goofy things for prizes, and you're going, "And you're paying me how much for this?" I mean, truly, when I say it was the best of times, as far as my radio experience, I think that was part of it, because it was just like I was a kid in a candy store, going, "This is cool." You know, you're just going out, having fun, and they pay you for that. And they pay you for that? That's crazy.

BROOKE HUNTER

I remember when I was hosting nights, Goldfinger came by the studio, and it happened to be my birthday. I remember turning around to pull some CD's or answer a phone call, and when I looked up, the drummer of Goldfinger had stripped down completely naked, stuck a drum stick up his ass, and the band proceeded to sing "Happy Birthday" to me, live on the air!

JON REENS

No Doubt at the Metro, I was the only P.A. (Promotions Assistant) there. It was freezing cold out; I got there early. "Just a Girl" was massive; it was their only hit at the time … and I was sitting in the balcony at the metro, and this line of girls comes in, and sits right in front of me. One of the girls turns around and goes, "I'm sorry, we're probably blocking your view," not knowing me from Adam. I'm like, "Oh, it doesn't matter. I'm here for work anyways." She's like, "Oh, what do you do?" I was talking to her, and I was sitting there thinking I was some badass because I worked for a radio station and here were these girls, and I was at the show for free. It was Adrian's girlfriend. So after the show I got invited downstairs because I was small talking with this girl, not knowing who she was, having fun with her friends; I bought them all a round of drinks. So she brings me back for the meet and greet. I had never been back in Metro's dressing room area; so, it was this little hallway with all these rooms. (Metro owner) Joe Shanahan was back there, a couple other people were back there, the label guys … they had run out of beer, so one of the Metro guys ran down to Smart Bar, grabbed more beer, and brought it up. I was sitting there, talking to somebody (when) Tony Kanal felt that it was appropriate to dump a beer over Adrian's head. Completely sold-out show. I think it was the first time they had sold out in Chicago, if my memory serves me correctly. So everybody was just kind of over the moon; so, it turned into this huge beer fight backstage.

Then, we did that toy drive (with No Doubt), probably six-seven months later, in the first floor of the Mart. As I'm standing there and all these guys are coming in. As Tony and Adrian walked in, they sort of stopped and looked and smiled, and just started laughing. I was, like, "Yeah, that was me" —not that they really remembered who I was or my name, or anything like that; it was just this experience that I had with these people, that was common, that they had as well.

SAMANTHA JAMES

The time I had spent with (Shanoon Hoon of Blind Melon) at Reading Fest, he was absolutely delightful. He wasn't a drug addict around me; he was drinking a Diet Coke and he was absolutely delightful. And I talked about that on the air. And his sister—you don't 100% know this was who she said it was—but Brian Peck took the phone call: his sister had heard me say that and she didn't want to go on the air or anything, she just wanted to call and say thank you for saying such nice things about her brother. That was kind of cool, because I think there were those moments where you actually realize you talk a lot of rubbish; so, it's kind of nice when you think you've actually said something, you know?

BROOKE HUNTER

As much as Q101 has been loved these past 10 or 15 years, when we first started on Q101, it was the cool station to hate. We were called sellouts: "I can't believe you're playing Pearl Jam, and Nirvana, and Soundgarden—people don't want to hear that on the radio. I can't believe you're doing that."

BRIAN "THE WHIPPING BOY" PARUCH

I remember feeling very ... like we were supposed to be ... I felt like every white person between the ages of 13-30, let's say, pretty much listened to us, and knew us. But I also felt like a lot of them hated us; I knew that from working promotions. Now, I'm pretty convinced it really didn't matter; it didn't really matter what the die-hards thought, because they weren't really the mainstream people of Chicagoland. But the die-hards who thought we were stealing their music and lifestyle and all this stuff, they were really, really, really, angry and vocal.

BROOKE HUNTER

I still remember 'introing' a Beck show at Metro and having some jackass whip a Miller Lite bottle cap at my face.

KEITH SGARIGLIA

I remember we'd brace ourselves: "Keith, you're giving away stickers after the Rancid show at Metro." And it was, like, "Oh really, am I?" You kind of knew what you were going to get … You knew when you were just going to get hammered.

SAMANTHA JAMES

I could be very confident sitting on my own in the studio. I got so nervous every time I had to go on stage and introduce. I can remember being at those shows, and you'd step out on stage: "I'm Samantha James from Q101," and there would be however many thousands of people cheering, but I could only see the one guy who was flipping me off saying, "You suck." That's all I could see.

STEVE FISHER

You wanted to wear stage armor, you really did. You wanted to wear some sort of steel armor, I should say. And it was—it could be scary. And I think the only way to overcome that was just to have fun with the crowd and not take it personally.

REY MENA

There were the fans, and there were the fans who didn't want to act like they were fans.

BRIAN "THE WHIPPING BOY" PARUCH

I remember some teenagey girl calling up, literally being in hysterics, crying, because we were ruining Marilyn Manson for her. Now that we were going to play it, all the kids were going to listen to it, and we were ruining it, and we were destroying it. She was literally crying; she was ordering me not to play it. The Nine Inch Nails people, the Marilyn Manson people, we weren't cool enough to be involved with their artists. The artist was never evil, by the way, in their eyes. It was always our fault. We stole the artists against their will.

KEVIN MANNO, *Air personality*

Q101 is responsible for getting me into music. I didn't care about music until I started listening to Q101, and then I won tickets to Jamboree. It was the year that Beck and Jamoriquai were there, was that 97? I won tickets that year; I was 14 years old. From then on, I listened all the time, like, all the time. When I was sitting at home watching TV, I would have it on because I didn't want to miss anything, and I would turn it up when the DJs were talking. And I remember just annoying the shit out of Brian "the Whipping Boy" over AOL instant messenger, sitting up in the morning and talking to him. If I saw him tomorrow, we know each other now, but I will never tell him to his face that I used to do that to him because it's so embarrassing. I knew that he hated me back then. I was just a young, huge, fan.

RYAN MANNO

It's all we listened to. (One) afternoon, we had a trampoline and we were out wrestling on the trampoline or something. I think it was Steve Fisher or Robert Chase who gave away Jamboree tickets for caller 101. Kevin scrambled for the phone while we were jumping on the trampoline, and he was caller 101; that was our first grown up concert, going to Q101 Jamboree the year that Erasure headlined, the Bosstones were there, Moby, Social Distortion. Really, it's all we listened to.

JEN JAMESON, *Air personality*

That was the era where it was still Goo Goo Dolls, Natalie Imbruglia, Chemical Brothers; it was sort of all over the place, which I think was when it was most fun. You could see a Sarah McLachlan sponsored show, and on the other hand, see Prodigy, Underworld …

BRIAN "THE WHIPPING BOY" PARUCH

I remember (interviewing Gavin Rossdale from Bush at Jamboree '97) — thinking like I was doing the Rolling Stone interview — I've got specific quotes from Trent Reznor ripping on Bush and how bad they were, and how he doesn't like them for this, that or the other reason. And I sort of asked him to respond to that stuff — I was, like, "How do you feel about Trent Reznor saying this?" or "How do you feel about this person saying this?" or whatever. I think I thought was, like, "I'm not letting this guy off of the hook." He smiled through it and answered the questions. Afterwards, he was very angry. He was cursing, saying, "Why didn't you ask me about things like people on this bill we like?" He was so angry. And then after, our photographer took pictures of me posing with (Rossdale and Bush guitarist Nigel Pulsford), and he's got this look on this face like, "I hate this. I don't want to be here and this person's an idiot."

17

CHAPTER TEN

Q101 broadcast from the 17th floor of the Merchandise Mart until January, 2001. The offices and studios were completely out of place on the restricted floor, which was home to wholesale dealers, designers, and well-heeled clientele.

CHRISTIAN "CAP" PEDERSEN, *Promotions Assistant*
The Merchandise Mart is a colossal building constructed by Marshall Field & Co. in 1930 and, at the time it was built, it was the biggest building in the world. It even had its own zip code. To get to Q101, you would have to walk inside this magnificent building and take one of the elevators up to the 17th floor. Once you got off, you would have to walk the equivalent of a city block down a dimly lit hallway, passing several showrooms of the most expensive, and sometimes obnoxious, furniture imaginable ... a mixture of the type of furniture you would find in your great grandmother's house and those you would find inside the lobby of the W (hotel).

Q101 was tucked in the southeast corner of the floor, visible by its silver and blue neon logo that shone through the glass doors that separated the lobby from the hallway. Q101 definitely did not belong in the Merchandise Mart; so, being tucked in the corner suited it—and the other tenants—just fine.

ZOLTAR, *Air personality*
The only building in the world that was bigger (than the Merchandise Mart) by square footage was the Pentagon. When I'd get off the elevator, I'd literally walk half a city block to the front door. I loved it, because I'd just go, "Yeah, I work

here." For me, I was just getting psyched out and pumped up and by the time I got to those glass doors, I had turned into Zoltar.

MARY SHUMINAS, *Music Director*

(The studio facility) was smaller; that's where Q101 as an alternative (station) was born. There were a lot of long hours there, and (it had) more of a family atmosphere. So I think there's a sentimental attachment to that facility. Had Q101 been born on the second floor (of the Merchandise Mart), we'd probably have the same sentiments about that facility.

BRIAN "THE WHIPPING BOY" PARUCH, *Air personality*

The (control) board was really old. It was older than whatever we had in Champaign. The buttons were old, blocky buttons and it had this big, ripped-up padding right where you put your elbows. It was an older board; it seemed a lot more clunky, I think.

SAMANTHA JAMES, *Air personality*

It wasn't a glamorous studio back then. You had all your CDs and your carts for your commercials. It was before the computer age.

STEVE FISHER, *Air personality*

Old. It was old. I mean our control board—there was nothing technologically advanced about our studio.

J LOVE, *Producer*

It was this dinky little studio that felt claustrophobic at times. On the left hand side, was a wall of 8-tracks, and even behind, because back in those days, we were playing things off cart. Denon CD players, kind of a weathered board which some might say had all the characteristics of great history behind it.

SAMANTHA JAMES

It was a bit of a tiny, funny little room.

BRIAN "THE WHIPPING BOY" PARUCH

We had the reel-to-reel recorder for phone calls. We sound stone age, but seriously, when we first started, we had a reel to reel, and seriously, we had to sit there and cut the phone calls up with a razor blade and a grease pencil. I still have the muscle memory in my mind of how to do that; I can feel my fingernail pressing down on the little tape to get that really good, seamless, edit.

TIM VIRGIN, *Air personality*

I remember on the 17th floor, I remember walking down that hallway, and you'd turn and look through those glass doors and see that fucking sign. That sign will be in my memory until the day I die. That sign meant so much to me, that neon sign. It was the coolest fucking logo sign I've ever seen in my life. That Q101 logo, to me, was one of the coolest radio logos ever ... It was so Nine Inch Nails; it was fucking awesome.

PHIL "TWITCH" GROSCH, *Promotions Assistant*

Coming into that 17th floor facility as an intern, and knowing that I wanted to be on the air, the air studio was a very magical place, even the placement of it, and the way it was set up. Everything else was very, "Oh, hey, it's office space, it's a bunch of cubes." But the doors to the air studio were two big, brown doors that would just open from the center. And you'd walk in, and the air studio was almost its own mini complex: main studio on the left, auxiliary studio on the right. You could walk through another set of doors to the quote-unquote newsroom, where the phone person would sit. The performance studio, another news studio, and they all kind of looked into one another through the windows. It was very ... people were nervous to go in there, especially as an intern or as a junior member of the staff. Even to open the doors and walk in, and hand Robert Chase a log or something, or a giveaway sheet was a fucking thrill. Because I just walked into the studio, and there's Robert on the air ... being that close to it was incredibly exhilarating and highly motivating.

"HERE'S A RING"

CHAPTER ELEVEN

Long before the shocks and reveals of reality television became pop culture clichés, two of Q101's air personalities played out one of their most private moments on the air, much to the surprise of the audience.

LANCE TAWZER, *Air Personality*
   (Samantha James) was there when we were taping the local shows, because they used to tape them during the day. She had these cute little pigtails in her hair. She was doing her mid-day show. I caught a glimpse of her through the glass; I didn't really meet her at that point, I don't think.

SAMANTHA JAMES, *Air personality*
   I remember him waving at me through the glass and I can remember thinking, "Oh God, why didn't I wash my hair today?"

LANCE TAWZER
   We were doing our show. We recorded the local show and maybe I met her in passing, I don't know. I met her in earnest the week we filled in for Robert, and Sam will attest, I had a girlfriend for nine years prior to that, and it was a steady kind of thing. As soon as I met her, I dropped my old girlfriend like a hot potato and pursued Sam with earnest.

SAMANTHA JAMES

It wasn't long after that that Robert Chase went on vacation and (Lance and Stoley) filled in for him, and that's when we started to get to know each other.

SUSAN BANACH, *Air personality*

Lance kind of liked Samantha, and Sam and I would go and do a lot of things together. One evening I was going back to her house because sometimes I would housesit for her … she invited Lance to do something, but I was there, so we all kind of went out together as three of us, as friends. I was sitting in her living room; she was living there on State Street. The evening was kind of going on, and it didn't take long for me to go, "Hmmm, I think this is a date, and I think it's time for me to go home." And they ended up getting married. There was a chemistry that was going on there where it wasn't just three friends hanging out anymore.

LANCE TAWZER

We were keeping it a secret. It was frowned upon, you know, to mess around among the staff. It wasn't until we got busted by Chuck Hillier that it became known in the radio station. Four months after we started dating, we got engaged. And we waited a year to get married. It was surreal, because for a time there, we felt like were living on top of the world.

SAMANTHA JAMES

We didn't tell Bill, and we didn't tell management, because we weren't really even sure whether they'd think it was cool or they'd think it was terrible, and we didn't know where it was going to lead, so why even go down that path? We actually got busted outside the Merchandise Mart. We went out after my shift. He had come to pick me up, and we had been holding hands crossing the road, and all of a sudden, there was Chuck Hillier and Rey Mena standing right there.

LANCE TAWZER

Within a year span, I broke up with somebody, met Sam, and within four months, decided we were going to get married. We brought it in to Bill, and said, "What do you think?" He said, "You've got to do it on the air."

SAMANTHA JAMES

Bill actually said he was more worried about if we broke up, and he said, "Who knows? Maybe that would make for really good radio. Maybe we could put you on in the mornings then."

LANCE TAWZER

She didn't know I was ready to pop the question or do anything like that. We talked about it for a couple of weeks, and (Bill) said, "Here's what we can do." He got the marketing people involved because he was sure that we could get, like, Gingiss to give me a tux, and all these other ways of turning this into a real bit.

I called her dad and asked for permission on the air. She comes in (the studio), and she kind of knows right away that something's going on, because all the lights are dimmed. My moment of crisis: I had all this coaching, all this getting me ready for it and I totally choked. I got all nervous, and I couldn't remember what anybody told me. I stretched my arm out and said, "Here's a ring." She got this nervous laugh and Stoley goes, "Here's a ring?" I said, "Oh, crap…will you marry me?" She said, "Yes," and then we all started yelling, and then, of course, Stoley comes back with, "That was fun. Here's some Pearl Jam." That was how the bit ended, if you can believe.

SAMANTHA JAMES

He actually never even really proposed; he said, "Here's a ring." He forgot actually to say, "Will you marry me?"

STOLEY, *Air Personality*

I mocked him. I remember now, I go, "Here's a ring?"

That is bad, though. First, you're going to do it on the radio, which is almost worse than doing it at a ball game on the Jumbotron or wearing a gorilla suit at a restaurant. Like, all of these are bad ideas. Then, not even to have your words planned out. You're broadcasting this to a sizable portion of the Chicago metropolitan area, and you don't even have a couple of sentences written down? You come up with "Here's a ring?" Come on.

SAMANTHA JAMES

It's one of those—it's got its pluses and minuses, being proposed to on the radio.

LANCE TAWZER

So the next thing we know, we're in *Billboard Magazine* for getting engaged on the air. The Radio Hall of Fame petitioned the station for the tapes to go into the archives. That trade journal, *FMQB*, did a "Women in Rock" CD for cancer. They had all these women DJs from all over the country, snippets of their shows, on this CD that was going to be distributed to all the radio stations; it was a trade CD. Sam's selection was our engagement, which mortified her, because that was not good radio and didn't really reflect what she was all about. That was kind of surreal.

CHUCK HILLIER, *General Manager*

All I can say about Lance and Stoley is, if you look at all of the morning shows that contributed to that landscape change, you would have to agree (on their show), if you were playing the game "which of these does not belong." I don't mean doesn't belong on the air; I mean does not belong in the same mainstream sort of morning-show host kind of stuff with all the other talent that was there.

BILL GAMBLE, *Program Director*

When you're getting up, going to work, the fundamentals (are), "Tell me the weather, tell me what time it is, give me the news." It's like, "Wait, I have to go to my job. I can't be listening to two guys talking about dope and masturbation."

CHUCK HILLIER

The most abrupt change that I recall was Lance and Stoley. It seemed to me to be the land speed record for in and out. The feeling there was, the audience hated them. Let's make a change in its most positive construction before we do further damage to ourselves. I don't honestly remember what their next rating book was, and it doesn't matter if it was only their first. These things take some time. We were never giving anybody, anybody, any time.

SAMANTHA JAMES

That's the thing I used to say to Lance, you know, enjoy your moment in the sun because I've seen so many morning shows come and go. I mean, it just seemed like every sixth Monday there would be a new morning show.

STOLEY

I don't remember ever forming thoughts about the future of a lot of stuff in those days. I was pretty young, and I was pretty much, like, "Whatever's cool with me. Doesn't everybody get a radio show at Q101?" I never thought, like, this could grow into something good. Even up to the point where after it was gone and then I lived in New York for a few years and then I looked back; I was, like, "You know, maybe I could've gotten a job in radio using my experience at Q101. Perhaps being the morning show host at a major market radio station was something you could use to work elsewhere. It never occurred to me; I was so not forward-minded. I was a very in-the-moment kind of guy.

LANCE TAWZER

The only reason why we ended up on the radio is that Carla liked our record, and liked us, and gave us the opportunity to come in and record in the studio, and be interviewed for the local show. That led to meeting Bill Gamble, which led to the radio shows, which led to all the great gigs, which led to the morning show and meeting my wife. I owe a heck of a lot to Carla.

STOLEY

It was really weird that we had managed as a band to get a record deal, but we didn't really have the force of a huge following until the radio stuff, and the radio stuff kind of came because we had something to talk about with the record deal. All of a sudden, we had all of the pieces to be like this big, hip thing in Chicago. What's surreal about it is that it didn't expand beyond Chicago. We were huge in Chicago, and then, nowhere else did anybody hear of a Lance and Stoley or Lupins.

LANCE TAWZER

I was glad that I had got to meet Sam, obviously. We look back at that and go, if nothing else, we have two kids, we have a 15-year marriage, we have this amazing story we can tell at cocktail parties to people who never heard of us. For quite a while there, there were a lot of kids that were disappointed that we weren't on the radio. They would come to our shows and send us emails, and that was very sweet for a while.

WENDY AND BILL

CHAPTER TWELVE

Of all the morning shows that went through Q101's air talent revolving door, there was a sense that perhaps the station had finally found its answer in former WLUP co-hosts Wendy Snyder and Bill Leff.

BILL LEFF, *Air Personality*
My goal was never to really get into radio, and I thought if I ever did get into the radio business, the only (station) that made sense was the Loop (WLUP). I didn't want to jump from station to station to station. Bill's offer was really interesting, of course, to do mornings, because I'm not a morning person, and although I like music, I didn't think I knew enough to be on an all-music format. But Bill said, "The personality that you and Wendy bring to the radio is exactly what we're looking for at Q101. And if you could do more than a month at a clip, that would help us out, because everybody else has only been on for a couple of days or weeks."

BILL GAMBLE, *Program Director*
Bill (Leff)'s a funny guy. And he's really good.

MARY SHUMINAS, *Music Director*
We never dedicated enough time to one (morning show). The one that I really think had potential was Wendy and Bill. Wendy Snyder and Bill Leff were doing talk at night at the Loop. To this day, I think Bill Leff is one of the funniest people around, and Wendy is very talented. I feel that they had more raw talent than

Eric and Kathy (WTMX morning show). The timing of that show was unfortunate because Bill (Gamble) left the station, and I don't feel the station really fully supported that show as far as promoting outside the station, like billboards or television. That was never done.

J LOVE, *Producer*

It was unique with Wendy and Bill because she was the lead and he was the second banana, but he was clearly the "funny" of the show, and then sometimes, she got painted to be the serious of the show.

WENDY SNYDER, *Air Personality*

Bill didn't have to work blue to be funny. He was always really cool. Working together, it was like the guy and girl "View." But surprisingly, I mainly had the guy views, and he mainly had the girl views.

JEN JAMESON, *Air personality*

Wendy and Bill had a pretty good dynamic back then.

BILL LEFF

I had to take a crash course on learning alternative music.

WENDY SNYDER

When we started at Q101, I was somewhat familiar with the music, but Bill had no idea. It was kind of great to get back to the music end of things.

BILL LEFF

At Q101, because the culture was more music-oriented, the audience had to be trained a little bit to understand what we were doing, and I think we had to be trained a little bit to understand what they wanted to listen to. So there was a period in the very beginning where there was a certain transition where the things that we did at the Loop and knew worked had to be reformatted a little bit to make them palatable to the listeners at Q101.

WENDY SNYDER

It was interesting. Bill is not a huge music lover. He likes music, but his musical tastes are quite different.

BILL LEFF

Barenaked Ladies. I had so much incredible fun when we did a concert with Barenaked Ladies. There's an action figure called Hugo, Man of 1000 Faces, which was an action figure out around 1970 or 1971, and it was a guy with no hair or facial features. It's actually a Billy Corgan figure, really. You could put eyebrows, or glasses, or a moustache, or hair on him. It's a rare thing to have a Hugo action figure. They had one up on their keyboard during a concert we did with them, and after the show, I came back to meet them, and I go, "Guys, as much fun as it is to meet you, I'm here to say hello to Hugo, Man of 1000 Faces." They all just gasped, and they went, "You know Hugo Man of 1000 Faces?" I go, "I'm the biggest Hugo fan there is." So we talked about Hugo Man of 1000 Faces for 20 minutes.

WENDY SNYDER

My biggest claim to fame: I learned a valuable lesson. Gwen Stefani was supposed to call at a certain time, and she never called. None of her people called. Maybe it was our fault, but Bill and I were, like, "Yeah, we've got Gwen Stefani on; she'll be calling up in, like, the next five minutes or so." An hour goes by and Gwen Stefani's manager calls and goes, "Oh, I'm so sorry we're late, but Gwen was in the hospital." We're like, "Oh my god, is everything okay?" "Oh yeah, yeah, yeah, she's fine. She was wondering if she could maybe go on now."

And it's, like, I don't know, five minutes before we're going off the air, and I probably did a snotty, stupid thing, but I said, "You know what? We've moved on, but thanks," which was so rude, but I mean they were rude for not calling us saying, "She was delayed." Someone could've blown in a call.

So there was this big show at Metro; this was when Bill and I just started there. Gwen Stefani gets up on stage, and she's, like, "Hey, Q101, what's going on? I just wanted to tell you that I was going to be on your morning show this morning, but they wouldn't put me on the air. I had a little problem, and they told me to forget it. So you know what I've got to say to you, Q101 morning show? Fuck. You."

BILL GAMBLE

They fit the radio station better over time. Both (were really talented, (and came from) different backgrounds. I'm not sure that you could – you know, we went the music route with Lance and Stoley, and then we said, "Okay, now we have to find some people that have the basic skills that you need in morning radio, and also, some good chemistry and talent." And, again, I think that's another show that, if it had been left on the radio, would've been very successful. And it was starting to see some growth, I think, before they pulled it off the air.

BILL LEFF

(Backstage) at one of the Twisted shows, they had like a "mess hall" with advertisers and people who worked at the station in sales and stuff.

It's Wendy and I and a couple of people from the show, Bill Gamble's there, and Chuck Hillier's there, and all these advertisers are there. And one of the advertisers comes up to me and Wendy, and he puts his arm around each of us and says, "I've gotta tell you guys something. You guys are the first show that my whole family can listen to together. I've got two teenagers, my wife and I, and we've never agreed on a radio show. Now all four of us listen to the show together and I love it." That, I thought, was the best compliment I'd ever had at the time.

Chuck Hillier doesn't say anything. But Gamble's looking at me, I'm looking at him, and the guy walks away. I said to Bill, "Wow, how about that, is that nice or what?" He goes, "That's not so nice." I go, "Why is that not nice?" He goes, "Take a look at Chuck. Chuck is grimacing." I go, "Why does that not make him happy?" He goes, "He doesn't want people in their 40s listening to the station. That's the last thing he wants." To this day, and I'm not great at math, I can't figure out why having two people listen to a station is better than having four people listen to a station. I've used a calculator, I've used an abacus, I've had people dress up like numbers and move in front of me, and it never adds up that way.

WENDY SNYDER

We felt like we could've fit in. There wasn't a lot of support from the station. I hate to sound unappreciative, but we didn't even get a Wendy and Bill t-shirt. We offered to pay for it. I'm serious.

BILL LEFF

We looked into the costs of everything from a billboard to our own t-shirts. Morning radio is very competitive, and we didn't feel like we were being supported properly, so Wendy said to me, "We should really look into doing this ourselves," and we did.

BRIAN "THE WHIPPING BOY" PARUCH, *Air personality*

I really thought (being the night DJ) was really like a "thing." I was excited about it; I thought I was doing great; I thought it was really fun, and it was fantastic. So when Bill (Gamble) asked me to go off of that, I was pretty upset about that, even though it was to do news on the morning show, the new morning show with Wendy and Bill. Ideally, in a perfect world, that would mean this show is going to be our signature show, it's going to be big, and you're going to be part of it. I think I took it as kind of a demotion; like, you're doing news, third fiddle to these two people. But I did it. It turned out to be fun in lots of ways and, looking back from the 2011 perspective, I think (it was) a pretty important thing for me to have done, because so much of my career since then has been in a similar mode to that.

BILL LEFF

It started out a little tenuous, because when I started, I looked at the roster of who was at Q101, and there was the Whipping Boy, there was Shark, and there was Zoltar, and everyone had a pretend name. I actually had in my contract—I'm not making this up—I must be referred to by either "Bill" or "Bill Leff," and nothing else, because I didn't want Bill Gamble to look at me one day and go, "Your name's Whipped Cream Head."

He said, "Whip" or "Whipping Boy" all the time, and I go, "Just for legitimacy, can we go with Brian?" He goes, "Oh no, people love him as Whip." I go, "Yeah, but I think we're starting fresh." I think Brian may have been upset with me a little bit, probably rightfully so.

BRIAN "THE WHIPPING BOY" PARUCH

No, I don't think I was upset about that, because it turned into Brian the Whipping Boy then. I was never really upset with them about anything. I think that I was sort of, at least for a while, upset with the situation, because I think I thought I was doing something really neat at night.

BILL LEFF

I looked at Brian as someone who was so incredibly knowledgeable in every single direction. He always reminded me of Bob Costas, in that there's not a single thing you can toss out there that he's not going to have not only some knowledge of, but be able to turn into something interesting. I have nothing but respect for him.

BRIAN "THE WHIPPING BOY" PARUCH

I have nothing but good things to say about him, too. That's really true. Like Robert Chase, (he's) legitimately the nicest guy who doesn't seem to have bad thoughts toward anyone, or doesn't seem to have any negative feelings toward anyone or anything, really. Bill on the air and off the air is just naturally funny, and funny in a professionally funny way. Not just like, funny, like, your cousin's funny. He's funny in a way that's like, that's a professional joke, and you just thought of it right now. And all the time. And never miss.

WENDY SNYDER

I feel like there's a reason my name is Wendy, because Bill Leff is like Peter Pan. He is like a man-child, a boy who has never grown up, the funniest guy I've ever met, the quickest guy I've ever met.

BILL LEFF

I think, mentally, I'm the same person I am now that I was when I was twelve.

BRIAN "THE WHIPPING BOY" PARUCH

I think I thought that Wendy, at least sometimes, almost didn't want to adapt to the station, as opposed to wanting the listeners kind of to adapt to them. Maybe in their minds, or maybe in Wendy's mind, she was adapting a lot to Q101, but I felt like it was more like, now a Q101 listener was supposed to adapt to a show from the Loop. Now I just see, she was doing her show the way she knew how, and was probably told by Bill, "Come and do your show. That's what we want." At the time, I think was, like, we, the station built this thing, we're trying to do this thing, and here's what it is, and here's some of the important people in the format, and all that. And it was just kind of, like, I wasn't sure what mattered anymore, because Wendy didn't treat that like it mattered. But again, that was also me being 23, thinking that I knew more than I did.

BILL LEFF

(For) most of the show when I worked with Wendy, she just made me laugh every day, and I'd like to think I made her laugh every day. It never felt like a job; it felt like the same kind of atmosphere you have at your high school lunch table, where you know it's going to be the highest part of the day. You're just going to have fun with whatever's going on in the news, or make fun of people at the next table, or make fun of yourselves. I knew that that's what I was going to get every day, working with her.

WENDY SNYDER

It's really hard to create chemistry. Bill and I totally had it.

BILL LEFF

Wendy is the perfect person to be on the radio, in that she does not have a filter. She just doesn't. Whatever happens to her in real life, she thinks that it's interesting enough for everybody to hear. And that's rare, you know? And I really learned the format of radio from Wendy. I was used to doing standup comedy, where a standup act had a certain ebb and flow and a build to it; you started strong, and you took it in one direction and then the next. In radio, because you're on for four hours, you have to pace it very differently, and I learned all that from Wendy. She knew how to do that; she had been doing it for ten years, and had been doing it really well.

MARY SHUMINAS

They could've been a somewhat hipper Eric and Kathy (morning hosts at WTMX, "The Mix"), and Q101, had they not gone so edgy, could have been a somewhat hipper version of the Mix.

BILL LEFF

In my head, our competition was Eric and Kathy who came on at the same time as we did. I always looked at it as a game of chicken. I used to say to Wendy, "One of the cars is going to steer away, and the other car is going to win the gold medal. If Q101 keeps us, and the Mix says, 'Okay we tried Eric and Kathy and it's not working,' we're going to be rich and successful for years to come and the station will flourish."

It ended up being exactly the opposite. It's funny because in the ratings, we would beat them for a month, and then they would beat us for a month, and that went on and on the whole time we went head to head with them. I think Mancow was beating both of us, but let's be honest; there are a lot more stupid people in the world than there are intelligent people.

BILL GAMBLE

Well, it's just—it's the way you build something. And, you know, Q101 was built on a foundation of music. And we had really excellent marketing, and we had great talent. But, we were never—and, it was, by design—you never want to be in a position, if you're in management, that talent can put a gun to your head. Why would you? And so, anyway, if Wendy and Bill stayed on Q101 and Q101 stayed its course, and grew with the audience and morphed into that hot AC radio station, Eric and Kathy—I mean, no. Bonneville was relentless. They're great broadcasters, but they saw an opportunity, and so it was Eric and Kathy, the "Birthday Game," marketing, and you know, not many people remember it, (but) they sat at a two-share for four or five years. Yeah, they kept plugging away, and plugging away, and plugging away. And, you know, they have talent. And, you know the combination of time and talent and marketing, usually works.

THE BREW AND BLOCK PARTY

CHAPTER THIRTEEN

It's pure coincidence that this particular chapter ended up at number 13. One of Q101's unluckiest (and ugliest) events in the 1990s was "The Brew and Block Party," an outdoor music festival starring INXS, Matchbox 20, and Meredith Brooks. The name summarized the general concept: for the price of a ticket, a fan gained admission and was provided with all the beer he or she could drink.

What could possibly go wrong with that idea?

REY MENA, *Marketing Director*

After doing all the Jamborees and Twisteds, we needed another event and we wanted to do something that was more street festival-like, but still big so that we could sell out, and have some buzz around it.

CHUCK HILLIER, *General Manager*

The deal there, pretty simply: this was where a member of our internal staff got the notion that we could have this "all you could drink" kind of thing. There had been a few citywide block parties where it was almost that way—not radio stations, but the City of Chicago, like a church would host one.

MARY SHUMINAS, *Music Director*

(The Brew and Block Party) was something initially we started putting together while Bill Gamble was still at the station. And one of our clients, I believe, had ties to U.I.C. We were able to secure a field on the U.I.C. property, and were going to put together something along the lines of the Old St. Pats Block Party, where it was bands and unlimited beer; so, that was the original concept. Somewhere late in the game, the university decided that they would not do unlimited alcohol. We had to change the event—this was really last minute. It was actually kind of a big deal getting INXS—they weren't at the peak of their career, but they were a huge band for playing (what) I guess you would consider a street festival. They were on tour or scheduled to be on tour; they were already booked at the Riviera, I believe. We needed them to cancel that date and fly here a day early in order to play our event. It was a huge deal to secure INXS for that event.

CHUCK HILLIER

There were some modest restrictions there, a lot of winking and nodding and that kind of stuff. Basically, someone on our staff introduced us to two guys that had City Hall connections.

No matter how it was structured or verbalized, the long and short of it was the idea that we could pull this off if we donated, let's say ten grand, to (the mayor's) favorite charity, or one of his favorite churches. It was unmistakable. We wouldn't just get on the air and say "free beer." There were city ordinances against it, but in a sense, what we fell victim to, and probably should've been a lot smarter about, no doubt, was the notion that here was this two-man team that was wired with City Hall, and they were going to smooth it through. They never misunderstood our intention of serving all the beer that people could drink. There was no misunderstanding. We had meeting after meeting and meeting...

REY MENA

It was basically 25 bucks. You (could) get in, and you could drink.

BRIAN "THE WHIPPING BOY" PARUCH, *Air personality*

I don't know if that's even legal, period. I'm pretty sure that doesn't sound right.

REY MENA

They freaked out because of the liability, and they pulled the plug on it.

CHUCK HILLIER

We got called down into a corporate boardroom on some stuffy floor, (and) got read the riot act. The two guys that were the intermediaries were sitting there at the end of the table, faces down, no eye contact, and basically, the attorney … was saying, "You can't do this; it's against the law." I would look at this guy who had veins popping out of his forehead and would shift my gaze and attention down to the other two guys who were offering nothing. There was no plaintive, "Yeah but..." There was no way that (it) wasn't a shakedown for more money.

REY MENA

At the 12th hour, the university decides to pull the plug on the "all you can drink" kind of deal.

CHUCK HILLIER

All I know is (that) it went from start to stop in about 90 seconds. We had bands booked, we had the facility booked, we had a charity on board that was going to be bringing in dozens and dozens and dozens of volunteers. It was a tragic moment.

REY MENA

So now, we're starting to talk about it slightly different on the air, but people had bought tickets, and all mayhem ensued that day because we had to go on air and say if (people) did not want that, if (they) were expecting it was going to be all you can drink, we would refund (their) money. So, we're sitting there with hordes of cash, having to refund money to what ended up being basically a bunch of drunks that wanted to come and drink for 25 bucks.

ALAN SIMKOWSKI, *Director of Market Development*

I wasn't in charge of managing those events, but we had a really big snafu there because we were originally selling tickets on an all-you-can-drink basis, but it was on a college campus. A lot of people screwed up and we had to give refunds back that night.

KEITH SGARIGLIA, *Promotions Assistant*

We kind of caught wind of the fact that, whatever happened with the liquor, we weren't going to do unlimited drinks, and there was going to be two tickets, if I remember right. That was when we said, "Okay, technically, we have to make sure people are allowed refunds if they want them since the terms have changed." It was determined that I was going to be working at this table.

We thought, "Come on, how many people are really going to ask for refunds? It's only ten dollars. They're going to come; they're still going to have the bands, and they still get two beers, which is their money's worth." It turned into this line around the block. I was there, I think, for most of the first two acts, giving away tickets. Finally, at some point, we cut it off. I think we just finally said, "Okay, that's enough." I just remember getting screamed at for hours from all these people who were pissed off at the station.

CHRISTIAN "CAP" PEDERSEN, *Promotions Assistant*

My job at this clusterfuck was to be the person who executed a partial refund of our listeners' money. The majority of our listeners complained bitterly, hurling a number of four letter expletives at me. I had to remind them that I, personally, was not the person responsible for what had happened and that I agreed that they had every right to be as angry as they were.

REY MENA

We went through what seemed like an eternity—it was probably the first two hours—of chaos, and then it became an unbelievable event.

CHUCK HILLIER

Not our finest moment. The one thing I can say is the people who demanded a refund were there for the free beer, not for a glorious night, some terrific music, and helping a charity, all of which still occurred. But yeah, that was pretty humiliating.

CHRISTIAN "CAP" PEDERSEN

At the time, it was the absolutely worst day I ever spent at Q101. However, I learned through that experience how to deal best with angry mobs of people and live to see another day.

BRIAN "THE WHIPPING BOY" PARUCH

All that I can say is the obvious, which is, "How could you possibly not know that in advance, before you sell something like that? Shouldn't the idea of unlimited beer raise so many red flags in anybody's mind? First of all, the fact that this probably isn't legal, and second of all, even if it is legal, don't you think it has the possibility of turning really, really, really, ugly?"

STEVE FISHER, *Air personality*

I think that was the last show that I introduced—I introduced INXS, and that was the last show that Michael Hutchence did in Chicago, I think. So I do remember that part. And I was a big—I was a *huge* INXS fan. And I actually remember meeting them at Lucky's after one of their Aragon shows. I just went up to Michael Hutchence, and I'm, like, "Hey, just to let you know—that song, 'Never Tear Us Apart,' that ballad," I go, "That was my college sweetheart and my song." He goes, "Great, great! Are you still together?" I'm like, "No no no, but still, 'Never Tear Us Apart' meant a lot to me at the time. You know, we've torn apart." But he was just a gracious guy.

ALAN SIMKOWSKI

That's one of the more sentimental moments, because it was the last time Michael Hutchence ever played in Chicago …

People joke to this day that Q101 killed Michael Hutchence because we made him play in a baseball field instead of a big stadium.

CHUCK HILLIER

It was an unforgettable sight, being on that U.I.C. baseball diamond that looked out over the city … it was like, son of a bitch, the one thing that we didn't control fucked us in the end.

"THERE'S SOME HOES IN THIS HOUSE"

CHAPTER FOURTEEN

Staffers were shocked in 1996 when Program Director Bill Gamble resigned to take a job to program and launch a new ABC-owned classic rock station (WXCD). Since Gamble was the person behind Q101's format transition and continued evolution, many were left to wonder who would take his place, and what changes that person would bring to the station.

BILL GAMBLE, *Program Director*

If you look at it from a musical point of view, we were into derivative bands by '96 or '97. Bands, like Sponge, that were Bush knockoffs, which were, on and on, etc. etc. etc. And, I know I've probably (said) this 20 or 30 times, but I remember being at, I think it was Twisted Christmas in '96, and I was walking around—I think it was at Rosemont—and I'm seeing 10-year-old kids wearing Q101 t-shirts that are going down past their knees. And it was sort of the telltale sign, to me, that, you know, this wave might be over.

MARY SHUMINAS, *Music Director*

The year Bill Gamble left, 1996—I believe he left in April or May—there was a period of about four months before they brought in the next Program Director, which was Alex Luke. For whatever reason, the consultants disregarded us for a while, and I felt I was able to do my own thing. I was Interim Program Director (and) James VanOsdol was Interim Music Director. As far as a lot of the programming, it was based on intuition, and not always based on research. In just a few months, we had some ratings success, so it was a great time for me because I got to do my own thing there.

ALEX LUKE, *Program Director*

(The station) had people that had interests and experience in alternative radio. Then you had people who had been at the station who predated the alternative format. You had people who were clearly strong personalities. It was a cast of characters, in a great way.

KEITH SGARIGLIA, *Promotions Assistant*

(Alex) was a genuine music fan; he liked music as much as anyone I know.

JEN JAMESON, *Air personality*

He was pretty chill … I got very little direction from Alex, which was not a bad thing, but I was very young and could've used it.

TIM VIRGIN, *Air personality*

Alex Luke was my program director when I lived in St. Louis. I was at the Loop at the time when Alex left the Point in St. Louis (and) got the Q101 gig. I was at the Loop at the worst time of the Loop's history, which was when they had that idea of making it an alternative hybrid, terrible radio station called the Planet.

STEVE FISHER, *Air personality*

I had actually heard Tim in St. Louis before he got hired at The Loop. And I thought, "Okay. Different. That's cool." Alex told me (replacing me with Tim Virgin) was a corporate decision. I later had an exit interview with Chuck Hillier, and Chuck said, "No it wasn't. We gave Alex full power to do whatever he wanted to do, and my kids hate me for it." But I think, at the end of the day, I mean, the reality is … honestly, it was time for me to move on. I'd been there five years. It's kind of like Menudo. When you hit 30, you kind of have to move on from Q101. I mean, you have to leave Q101 when you hit 30. I mean, on-air, you should, right? You should either go into management or something else. Or sales. And I wasn't about to go into sales. And they certainly wouldn't take me in management. So, I was just, like, it was time for me to move on. I mean, it really was. And looking back, I will say this, the day that I got fired, after I got off the air, Alex just said, "We're letting you go." And I just said, "Why?" And he said, "Oh, it was a corporate decision." And he even went on to say, "I told them, 'Look, no one works harder than Steve does at prep and this and that.' And I've really seen how you're really trying hard." And, you know, at the end of the day, I mean, it is what it is. It was his decision. He was the boss. I get it.

ROBERT CHASE, *Air personality*

(Tim Virgin had a) very natural talent. Things came very easily to him, a guy who in the moment thinks very quickly on his feet and doesn't necessarily have to do any preparation or anything. (He) kinda rolls with it in the moment, knows his music well and can hang with the band and hang with a crowd of listeners, too, and be just as comfortable doing both.

KEITH SGARIGLIA

I think it was an Oasis/Cornershop show. James, Mary, Brian (the Whipping Boy), … and I (were) all sitting together, kind of against the hockey boards that were up for those shows. (Alex) and Tim Virgin came in and they had their lanyards on, but they had a stack of what seemed like 20 laminates on their lanyards. And we all just kind of looked at them, like, what was that about? … These guys came in wearing their entire collection, and Brian looked at them and he was like, "You guys going to all those shows tonight?"

BRIAN "THE WHIPPING BOY" PARUCH, *Air personality*

I remember them showing up with like ten laminates apiece, you know like a person who has a lot of stamps on their passport, or a person who has a lot of stickers on their suitcase, from all the places they've been. That's how they showed up at some show. I don't remember exactly what I said.

ALEX LUKE

One of the things I was proudest of, and we did it for a short while, (was) Rodman Radio, where we brought in Dennis Rodman on Friday afternoons on Tim's show, and we made it into a party on the radio. It sort of epitomized everything that I really felt like we should be striving for. It was current and relevant because it touched Dennis Rodman and the Chicago Bulls, which were the biggest things anyone could imagine at the time. It sounded very real and very vibrant … and we did it in a very high profile, very prime-time slot. The fact that we could pull that together on a whim and give him some airtime—you know, Dennis Rodman comes in with Carmen Electra, and he and Tim are messing around at 5 o'clock on a Friday afternoon—made for great radio.

TIM VIRGIN

I met Dennis's friends, Marco and Joey, who were guys that were always kind of hanging around the city; we always bumped into each other. I went to a Bulls game and I had a purple Mohawk; I was wearing a black suit and a Mohawk. Marco was talking to me, and (when) Marco went down at halftime, I guess Dennis asked Marco, "Who the fuck is that guy?" Marco's, like, "Oh that's my boy, Tim, do you want to meet him?" And then Marco came up and got me, and after the game, took me down to the court. I met Dennis, and the first thing Dennis says, he starts talking about Pearl Jam, like, "Dude, that's my favorite band." I'm, like, "Dude, you should come on my show." That's what all DJs say, right? And he's, like, "Okay, I'll do that," and I'm like, "Yeah, bullshit, whatever." I remember telling Alex, and then one day I was on the air and the phone rings and it was Marco. He's, like, "Hey, I'm with Dennis; we want to be on the air right now." It's like a Wednesday afternoon. And I'm, like, "You can't just call and be, like, hey let's be on the radio." I remember going down to Alex's office. I say, "Guess who just called me? Rodman and Marco, and they want to come. They're with Carmen Electra or somebody, and they just want to come and be on the radio." Alex was like, "You should do it; you should put them on." This was the brilliant thing about Alex; Alex was one of those Program Directors who (thought that) sometimes breaking the rules or bad radio could go farther than having Dennis Rodman on in a way that was all laid out and stuff, like having Dennis Rodman on and not knowing what the fuck he might say. He was still a part of the '97 Bulls, and he was on Q101; he wasn't on any other fucking radio station. Dennis came on our show, not because of sports, but because of Pearl Jam and because he liked Live. He couldn't even say Yellow Ledbetter, but it was his favorite song.

ALAN SIMKOWSKI, *Director of Market Development*

I think the biggest celebrity moment ever was really when, with about an hour's notice, we got confirmation that Dennis wanted to come in and do Tim's show.

J LOVE, *Producer*

Mike Bratton and I had plans to go out, drink, grab some dinner, and have fun. Turns out that Tim Virgin was having Rodman in the studio on one of the now-famous, Rodman radio days. He was in the studio and it ended up that they wanted a keyboard in the studio for whatever reason, I don't even know why. It ended up that Mike had a keyboard, or a synthesizer, in his studio; so, he's, like, "Hey, we are going to go out, but we're not going to go out. We're going to do this thing: Rodman Radio, whatever." I helped him carry this keyboard into the studio on the 17th floor, and I basically was getting waters, whatever they needed. It was the classic intern role for a few random hours, totally unscheduled and kind of awkward. I think it was somewhat determined on the whim of Dennis Rodman. That was my first time standing in that studio, and I thought, "You know what? I like this studio."

TIM VIRGIN

I remember Dennis's only request was a (music)bed coming in and out of when we went into commercials and Dennis's bed choice was "There's Some Hoes in This House."

ALAN SIMKOWSKI

It was the first time; it was within days of Dennis and Carmen Electra announcing they were dating. They were in the studio, and it was Brian, it was Dennis. They were playing music, drinking beer, talking about all kinds of stuff. Dennis had a little bit of an entourage at that point.

TIM VIRGIN

Eventually, it turned into anytime Dennis felt like coming on, we would go into "Rodman Radio" mode … My vision for it was I was kind of in the studio, but Dennis was just going to do whatever he did, and be a DJ. That was Alex's thing: "Let him DJ, let him take it over. If it fucks up, that's amazing.

BROOKE HUNTER, *Air personality*

Alex didn't fire me; he actually demoted me. He fired Zoltar, and he moved Zoltar and Steve Fisher. He put Tim Virgin on in afternoons; he moved James (VanOsdol) to nights and he was going to move me to overnights. I flipped out and went to Chuck and I told Chuck … Alex said to me, "Give it a day or two to think about it." I said, "Okay, I'll think about it. " I turned around and I went to Chuck Hillier, who was our GM at the time, and I said, "This is not acceptable. I will not do this. I'm not going to work overnights; I'm not going to turn my life upside down." That's when I called Barry James (WTMX) and got hired for nights over at the Mix. The funny thing was, Chuck never told Alex that I quit, because back then we had non-competes. Chuck let me out of my non-compete and told me that I could go work for another station in town, so I took the night job. James was on at night; nobody ever showed up for the overnight shift. James called (Alex) and said, "Brooke's not here," and Alex then called Chuck, and Chuck said, "Well, she quit."

I did one (overnight shift); that's the only time I smoked pot in the studio, because I took a big bag of weed in with me that night. Bill and Wendy were doing mornings, and I sat in the studio all night long and smoked pot to the point where I couldn't even talk because my voice was so hoarse. I know that Wendy and Bill came in, and the whole studio reeked of pot.

MIKE BRATTON, *Creative Director*

Brooke was really awesome. She was probably one of my favorite people to shoot the shit with. … She had worse than a sailor mouth, and she was quintessential Chicago.

ALEX LUKE

In the late 90s, you basically had the format fragmenting. Some alternative stations were becoming what was called Modern AC at the time, so you had stations that were building their music and their sound around artists like Third Eye Blind, Gin Blossoms … artists that actually now sound incredibly non-alternative in the context of 2011. But at the time, a lot of what they were doing musically was very different from what was on the radio. And they were new bands that didn't have the baggage of being a rock band, or a pop band or of being pigeonholed into a certain slot. Some of the alternative stations were going heavily in that direction; other alternative stations were going off in a rock direction, which is what I had done with the Point in St. Louis. We had taken it very heavily into the rock arena …

BRIAN "THE WHIPPING BOY" PARUCH

When did we start playing Creed? Whenever that happened, I think I thought we went in the wrong direction.

Whenever we started playing Creed very, very, regularly, we played the song "Torn." Robert played that song , and that song might not sound weird now, but at the time it just sounded way too "rock" for what Q101 had been—way too "Rock 103.5," in the parlance of the day. Robert came on the air and actually said, "Q101, that was … that was crap." That song is even worse than "My Own Prison!" It breaks down and goes to this really cheesy, big, giant, chunky, chug chug chug chug chug chug chug chug … chug chug chug chug chug chug chug chug. … "That was, dot-dot-dot, that was crap!" (laughs)

ALEX LUKE

If pop radio is sort of sugary and fake, canned, and preconceived, alternative radio should be vibrant and real, and edgy. It should be giving you something that you're not expecting every time you turn it on.

THE COW IS NOW

CHAPTER FIFTEEN

Wendy and Bill weren't an obvious fit for Q101 when Bill Gamble put them on the air to replace Lance and Stoley. By early 1998, however, the show had started to gain ratings traction with Q101's initially skeptical audience, as well as with the station's behind-the-scenes doubters.

Regardless, the duo's ratings continued to pale in comparison to Mancow Muller's morning numbers at Rock 103.5. Mancow consistently ruled the coveted and hard-to-earn men 18-34 demographic, from ratings book to ratings book.

BILL LEFF, *Air Personality*

We were told almost every day, in post-show meetings, that Mancow was ruining radio, that he was the worst thing to happen to Chicago radio, maybe to radio on the whole, that he was making radio stupid. I agreed with that. I thought he was making radio stupid. Then, the joke ended up being on us, because he replaced us.

WENDY SNYDER, *Air Personality*

We had just gotten off the air. (Chuck Hillier) said, "I just wanted to let you guys know that we're not going to renew the last year of your contract. We'd be willing to give you a last show if you want."

BILL LEFF

There was never a minute where I felt that Chuck Hillier wanted us to be in the fold over there. Whatever we would ask for, as we were asking, we knew the answer was going to be "No." I always felt like a kid asking, "Is it okay if the Cubs come over and have dinner with us?" You know, as you're asking it, your mom and dad are going to go, "Yeah, we'll see if we can make that work." I always felt like whatever we would ask, big or small, they would pull out a prepared statement and read to us why we couldn't have what we wanted to have.

WENDY SNYDER

Part of us knew. There really wasn't a lot of support.

BILL LEFF

I thought we'd be replaced by somebody. I didn't think we'd be replaced by Mancow.

WENDY SNYDER

We were able to come in the next day to do a show; that was the first and only fight Bill Leff and I had. I said, "You can do what you want, Bill, but I'm not going to burn a bridge. I want to keep doing what we're doing in Chicago radio." Bill's like, "You're not the boss of me."

BILL LEFF

At the time, I really felt that the station not only sold us out, but they sold out all their listeners because they were representing the station as being something, and that something was cool, underlined 1000 times. At that time, Q101 was the coolest station in the city; nothing came in second place. It was just dripping cool. To go the route they went with Mancow, I thought, was saying to every one of their core listeners, "We fooled you. We wanted you to think we were cool, but look what we did instead."

I talked about that on the air. I said, "You're all being ripped off. Every one of you is being ripped off." I said, "They're selling you short. They're going with the least common denominator. They want you to be dumber. I hope you don't follow that route."

KEITH SGARIGLIA, *Promotions Assistant*

I thought it was interesting that they left Wendy and Bill on the air after they fired them, for one show. I remember manning the email inbox during those periods, getting the feedback. You always felt like everyone in the city was listening and pissed off that they made that change. I think everyone wanted some continuity there. None of the shows was great, I don't think. I think some were better than others, certainly, but I don't think we ever hit a home run.

CHRISTIAN "CAP" PEDERSEN, *Promotions Assistant*

I heard Mancow's Morning Madhouse was going to be replacing Wendy and Bill while I was listening to Wendy and Bill's last show on the Metra on my way to work … I was really surprised Wendy and Bill were allowed to go on the air to thank their listeners for being loyal and to vent about the changes that were to come after they were told they were going to be let go.

BILL LEFF

Wendy's probably the more professional of the two of us. She said it was their decision; we have to abide by it. I didn't see it that way. I didn't want to abide by it. I felt everyone was getting cheated in it.

I didn't care. I felt in two and a half years, we had cultivated an honesty and a bond with the listeners. (When) we started out, it was very rocky. I don't think that most of the listeners wanted us to be their morning show hosts. We did something at Taste of Chicago; we just showed up and we got booed on stage. After ten years of standup, no matter what, I'd never been booed anywhere for any reason. They announced our names, and that we were the new morning show on Q101, and we got booed just for being there. I said to Wendy, "Let's see how we can turn this into a positive."

As time went on, and we got to know the audience, and the audience got to know us. We developed a nice friendship with the audience, I felt. They just pulled the plug too early, and in the most weird dimension imaginable.

STEVE FISHER, *Air personality*

If they had been given a chance, they would have been Eric and Kathy, fifteen years later. If Q101 had stayed the course, and just stayed the course, they would be as huge as Eric and Kathy. Because, I think, that Bill Leff is a very funny man. And I think Wendy is a tough Chicago chick. And that's a great combination … I really think that, you take that model, and nothing against Eric and Kathy, they're very successful and I give them all the credit in the world, but to me, Wendy and Bill are Chicago. Eric and Kathy are Naperville. Does that make sense?

WENDY SNYDER

I feel like we could've really done something. It was a great time then. I really had a blast working at Q101. I got back into music.

BILL LEFF

I wish it had lasted a lot longer.

BRIAN "THE WHIPPING BOY" PARUCH, *Air personality*

I think we all found out from the (Robert) Feder column (in the *Chicago Sun-Times*) that Mancow was actually happening. I think it was on a Sunday, but I think it was like the front page of the *Sun-Times*.

DAVE RICHARDS, *Program Director: Rock 103.5*

At the time, Cow was still in his prime. He left the Rock at the top. He walked away. Emmis did some great marketing for him around town, and in one (ratings) book, he was right back up there again. It wasn't really a matter of, you've gotta change your gig. With any great show like that, you know what you're buying. You're not purchasing this tremendous talent to make it different.

CHUCK HILLIER, *General Manager*

Here's the dirty little secret -- nobody really listened to (Rock 103.5) after 10 o'clock. Their ratings for the rest of the day were abysmal. So they had this gigundo morning show--they had the opposite problem of Q101—they had this gigundo morning show and nothing to speak of after 10 a.m. And for some quarters, especially given where we were, we weren't plugging Mancow into a 25-49, female-leaning adult alternative radio station. The music had shifted dramatically over a period of time. The decision was made to bring Mancow aboard and, for whatever it represented, I knew it would bring instant cohesion with the male-based audience that we now had, and it did. But I also remember sharing with corporate, "Look, no problem that you want to do this, but I would be planning right now for a new set of call letters, and logo and format for when the inevitable happens. You're probably going to wind up firing him; he may wind up going somewhere else for a ton more money. In any event, he's going to be gone, and this radio station is going to be built on the back of Mancow, not the music. If Mancow leaves for whatever reason, Q101 will simply not survive in the same way, shape, or form."

It might've taken a little longer than I thought.

ALAN SIMKOWSKI, *Director of Market Development*
(Hiring Mancow) was one of the major coups in radio.

PHIL "TWITCH" GROSCH, *Promotions Assistant*
At this point, anyone who wasn't on the sales staff was not thrilled about the Mancow show coming over, because we all knew that would be the end of Q101 as everyone had ever known it to be up until that moment in time. Before that, there was Rock 103.5 and there was Q101. Mancow was Rock 103.5, and Q101 was alternative, and these two stations were at war. I mean at war. So to bring Mancow over to Q101, from a product point of view, was fucking heresy. How could we bring the enemy's general over to run our radio station now, or to be our marquee talent?

MIDGE "DJ LUVCHEEZ" RIPOLI, *Technical Producer (Mancow's Morning Madhouse)*
(Mancow) made fun of Queer 101 for years before he went over there. And once we went over there … how do you call everybody all the names under the sun, and then the next day, you're working alongside them?

MARY SHUMINAS, *Music Director*
It was exciting because this was something that the station was actually spending money on. He was a huge deal, and it was exciting for sure. But at the same time, maybe not an ideal fit for the music, but I think people bought into it.

ALEX LUKE, *Program Director*
Coming from St Louis, having had a lot of success at the Point (KPNT), programming it as a rock station, and a heavy rock station … I had seen these rock records live side by side with the heavy alternative rock records and work. Honestly, when we were discussing bringing Mancow on board, when we had the conversations with Rick and Chuck, part of the reason it was such an obvious move to me was (that) it gave the station real focus and real continuity. We could be edgy and rock-based with Mancow and own a position in the market that no one else could own. It was really clear that musically, there were a handful of

Metallica records that could live adjacent to Foo Fighters, Korn, and Nirvana, and sound completely contemporary at that time. And I think you kind of had to go through the catalog. We went track by track. I think one thing that I did, and I think Bill Gamble did before me, was when Q101 did something, we did it right and we didn't mess around. It was definitely one of those. I think radio, good radio, is all about being a very specific thing and all about super-serving a very specific segment of your audience, and being a very specific tangible thing for your audience. Whenever we would make a decision to go in a rock direction, or play records heavily, or pull in a giant personality, it was, like, if we're going to do it, let's be extreme about it. Let's bring in the biggest personality we can, or let's play records as much as we can, or let's go as far in a rock direction as we can, and I think it served Q101 well.

CHUCK HILLIER

Of course, we were losing our females with every shift adjustment that we made. What happened when we brought Mancow on board? We lost the last six that were left. Where did they go? (The Mix.)

BILL GAMBLE, *Program Director*

I'm not sure I wouldn't have hired Mancow, too, 'cause, you know, someone said, "Man, I can have a seven share in morning drive plus what I've got now?" Yeah, I don't know anybody that says no to that.

BRIAN "THE WHIPPING BOY" PARUCH, *Air personality*

When Wendy and Bill left, they kept me to play music in the ensuing period until Mancow was ready to start. So I got to use the really old, decrepit production studio to do the quote-unquote morning show. (Mancow's show) did their syndicated show out of the other room before they were able to start on Q101. Why would that be? I don't know, but they did. Turd would come in a couple times while I was playing songs, like, "Oh yeah, sorry, the Whipping Boy show over here. You know Mancow's coming on in two weeks, and you know you're going to be out of here right?"

J LOVE, *Producer*

I would go back and forth with Rock 103.5 because there was this non-compete period of time where the Mancow show could broadcast in Chicago to the national affiliates, but not to Chicago; so, that's where I met Jim Jesus (Jim Lynam), and people from the Rock 103.5 world.

JIM "JESUS" LYNAM, *Producer (Mancow's Morning Madhouse)*

Before we got there, Q101 was very much a music kind of station. The morning shows there were more a part of what Q101 was all about, whether it was Lance and Stoley, or Wendy and Bill, or whatever. They morphed into the fabric of Q101, whereas with Mancow, you're putting a bull in a china shop. It's a juggernaut. It's its own entity. I don't think that was something Q101 was used to. I enjoyed it when I first got there, because it was kinda cool, but Mancow was pressure-filled, and I think that pressure spilled all over the entire building. We were intense. I don't know if Q101 before that was so intense with, "We've gotta win," you know what I mean?

CHUCK HILLIER

The first day on the air, the entire sales staff and many other people from the radio station had Mancow t-shirts and were hidden in the dark lobby of the radio station; so, when he showed up at whatever it was, slightly after 4:00, 4:30, people really turned out to give him a welcome. It was kind of a neat moment, really.

REY MENA, *Marketing Director*

There were such antagonistic relationships at the time between Rock 103.5 and Q101 promotionally, and on air they would call us "Queer 101" … I said we can't have him just show up, have him roaming the halls, and have this weird feeling between him calling everybody there "Queer 101," and the whole thing. So, I was, like, "Okay, we're all now one family, and we have to change that." I convinced everybody - Sales, Marketing, everybody - to show up, 'cause they were going to walk into the studios for the first time at about 4 o'clock, 4:30, something like that, to go on the air for the first day. I talked everybody into showing up there so we could create basically a tunnel of people cheering them on, as they walked off the elevator and then walked into Q101 for the first time. And then we all had shirts that said, "The Cow is Now". It really broke the ice, because when we did that, it was, like, "Okay, whatever battles ensued before are left in the past, but now you're part of Q101."

PHIL "TWITCH" GROSCH

Here we all are. Some of us had to work late into the evening before. We're waiting in the lobby for the Mancow show to start coming down the hallway. Sure enough, down the hallway, (came) Mancow, his producer at the time, and some of the staff members.

Then he just walked right in and did the show.

KEITH SGARIGLIA

We came in, welcomed him in the lobby. For a guy who had his guard up, he seemed genuinely stunned and kind of touched that we would all do that. I don't think when they hired him ... I don't think any of us thought it was bad. We had research that supported that people who listen to him in the morning listen to us the rest of the day. I think we all thought, "All right, the one thing we've been missing is a morning show, and he's huge. So that can only be good." Most of us were on board.

JEFF DELGADO, *Promotions Assistant*

At first, there was some hostility since they had always been considered the enemy and we had sweet Wendy and Bill.

ROBERT CHASE, *Air personality*

I remember (Mancow) was clearly the competition when I was doing mornings. I didn't get to listen to him much, so I didn't really know his spiel. We certainly heard a lot about him. He'd been on the board as a possible morning person at some point, if it could be worked out.

What I remember about him arriving is there was a call out to the entire staff - sales people, everyone - to be at the station to greet him that first day. I don't remember if it was mandatory, but it was pretty much stated that everybody needed to be there to greet him. I said, "To hell with that. I'm sorry. I've been here a good chunk of years, and I'm going to be on after him. I'll see him soon enough. I'm not coming in at 4 a.m. to welcome him; I'm just not." He told me later that he actually appreciated the fact that I didn't come in. He thought that was a good thing.

MIKE ENGLEBRECHT, *Promotions Assistant*

There were maybe 20, 25 people there. I remember everybody standing there waiting. I believe we were all there wearing "The Cow is Now" shirts. And we had that long hallway, so you see these two figures—I had never seen the guy before in my life—and here comes these two figures from the far end of the hallway; it turned out to be Cow and one of his producers.

In they walked, everybody clapped. I don't remember what he said. I think he was fairly modest, almost embarrassed about it. They darted into the studio, and that was his first day.

AL ROKER, JR., *Air Personality (Mancow's Morning Madhouse)*

(That) kind of was a detriment, because, in the long run, you kind of had the feeling that our shit didn't stink, and so, I think we tried to get away with as much as we possibly could. (Laughs) And we did! We did for a long time. But, I found everybody to be great. I found the culture to be very relaxing. Everyone was really friendly; no one was stuck up. 'Cause I didn't know what to expect. I'm like, "Am I gonna be surrounded by a bunch of 20-something know-it-alls?" You know what I mean? And I never got that feeling.

The show name of "Mancow's Morning Madhouse" was fairly self-descriptive. As host, Mancow Muller served as the warden for a colorful, boundary-pushing, group of personalities, including Turd (stunts and commentary), Al Roker, Jr. (sports), Freak (traffic), DJ Luvcheez (technical producer), and Jim "Jesus" Lynam (producer).

BRIAN "THE WHIPPING BOY" PARUCH

When Mancow started, they wouldn't let Freak come (over to Q101), so they needed a traffic guy; so, he shoved me in there as part of the one billion people in the studio. Every day was just a taunt fest on the air about how I was going to be out of a job in November: "36 more days until Freak comes!" I remember thinking at the time, "Wow, they really like this Freak guy." And then once Freak really came, you realized that all (Mancow) ever did was make fun of him until he couldn't take it anymore, too. Since he was not there, he was this beloved character who they couldn't wait to get back from captivity from the other station.

FREAK, *Air Personality (Mancow's Morning Madhouse)*

We were not under contract with Mancow; we were under contract with RCX, and so when Cow left, they let everybody go, but they hung on to me. I became the grunt. I did all the promotions. I was doing Radio Anarchy at night, and I stayed there until they pushed a button to become "Mega" 103.5, and when they changed the call letters, that pretty much killed my contract; so, I was free. Cow didn't have to take me, but he did. He took me back.

BRIAN "THE WHIPPING BOY" PARUCH

Freak eventually came, but they kept me because Mancow had figured out this little niche—making fun of me was too much fun to give up.

KEITH SGARIGLIA

Nothing changed the culture as drastically as Mancow's arrival did.

JEN JAMESON, *Air personality*

The culture shift happened with Mancow.

MIKE BRATTON, *Creative Director*

When they came in, I was like this is fucking crazy. I don't want to be here anymore.

ROBERT CHASE

I was all for us doing well in the ratings and having a strong morning show after all the things we had done and tried. I didn't think it was a bad idea, but I knew there was going to be a cultural shift coming from the station. They were a hard rock station, so what did that mean to us? I remember having the conversation about, are we going to play Metallica or no? The whole crew came in and they were very impactful, high energy, talented people, opinionated, and partiers, and all that kind of vibe. And we were a pretty straightforward, kind of mellow group up until that point. So, here came the big shake up. Following him was a curious thing, because I was supposed to go on at 10 o'clock, but rarely was that the case. I'd be there at 9 o'clock, ready to go on at 10, but rarely did I go on before 10:30. I remember there were some days pushing noon before I went on. I was told just to sit and wait. Then I had to catch up on all the spot breaks that weren't to air while he was on the air because of advertisers who did not want to be included on his show. So, all the spots during the 10 and 11 o'clock hour that he would not run, I had to make up; so, that was great radio.

AL ROKER, JR.

Cowboy Ray was a mentally challenged 45 year-old man, who had been a special guest on the show many, many times, and, of course, always wanted to get laid. It was literally our first week on the air on Q101. So, Mancow has this friend go out and get a hooker.

BRIAN "THE WHIPPING BOY" PARUCH

Cowboy Ray and a prostitute from Cabrini Green. I didn't think I was ever going to forget it at the time, and I haven't. (Mancow) told (Bob Noxious), who was a real character, to go find Cowboy Ray a hooker.

AL ROKER, JR.

Yes, Bob Noxious. Go out and get a hooker for Cowboy Ray. He comes back with this African-American hooker (who) comes in the studio with Ray, and then starts to (long pause) pleasure Ray in the studio.

BRIAN "THE WHIPPING BOY" PARUCH

He brought back some lady who had four teeth missing; I think she was like a crack hooker, literally. Cowboy Ray was a new concept to me. One thing led to another, and his pants were off and there was some kind of a release that happened, right there in the studio on the equipment, and there was blood...

AL ROKER, JR.

Cowboy Ray climaxes in the studio. And I'm about five feet away from this, and I'm going—I'm kind of, I'm shaking my head and closing my eyes, and thinking to myself, "This can't. Get. Any. Worse." And as soon as that thought leaves my head, the hooker goes, "Um, excuse me, he's bleeding." And that was our first week on the air.

MIDGE "DJ LUVCHEEZ" RIPOLI

She pokes her head up and says, "It's bleeding." We were all like, "What is?" From that point on, everybody started getting grossed out and screaming.

BRIAN "THE WHIPPING BOY" PARUCH

And the crack hooker, I'm pretty sure, ran from (Cowboy Ray). We all ran. I remember running to the door, and I don't know what I said, but I could have seriously said, "I don't ever want to see this room, or this studio again. I don't need this job that bad."

J LOVE

(Brian) stayed, and he probably witnessed worse.

BRIAN "THE WHIPPING BOY" PARUCH

The sort of bottom line of it was this crack hooker from Cabrini Green was repulsed by him. Can you imagine the things she's seen or done? And she was taken aback. "Wow, this is crazy." She's saying, "This is too crazy for me."

J LOVE

(Mancow) and I were on the board side of things, and everyone else would be on the microphone side of things. It's ingrained in my brain, looking up, and you only see shoulders and heads at this point, from our perspective, but the crack whore just kind of looks up and looks at us, Mancow and I, and I believe her quote was something like, "He's bleeding." He was just so raw in that sensitive area, Cowboy Ray, God bless his soul, or God rest his soul ... she attempted to masturbate him but it went directly to blood. It wasn't blood from the inside out; it was blood on the outside.

CHUCK HILLIER

I remember phoning in, and apparently, they had pranked Cowboy Ray into some kind of activity, which resulted in him spilling his seed in the studio.

I got back in, and we've got the first of zillions of face-to-face meetings, and I said, "Seriously now ... you wait until I'm out of town for the first time to do this?" And he goes, "Yeah."

MIKE BRATTON

Unfortunately, I remember the Cowboy Ray masturbation story, typically in my cold-sweat dreams. I also remember another morning where I, for the first time in my life, really seriously contemplated a sexual harassment lawsuit against the company.

Mancow's crew grabbed me by what felt like gunpoint … drug me into one of the smaller little studios in the back, near the on-air studio … to observe with probably 15 or 20 other people, most of whom were unpaid female interns … a woman stuffing a variety of very large objects in her asshole. And we were instructed to "ooh" and "ah" and gasp, and make plenty of noises because a woman shoving things into her asshole … was not really great radio. But the sounds of disgust that everyone was making around it made great radio, apparently. The one vivid, vivid, thing that is now suddenly coming back to me as I remember that horrific moment was the smell in the room, because you can't insert and move in and out large cucumbers and (a) fist without a pretty, pretty, enormous amount of lubricant.

MIKE ENGLEBRECHT

I remember Cow grabbing a handful of people and me, and herding us into one of the side studios where visiting bands usually would play. When I walked in, there was a woman in the studio, completely nude, placing various objects into her … back door. At one point, she squatted down like a catcher on top of this enormous rubber object … I'm pretty sure it was called a "King Kong" something-or-other, so you can imagine how big this thing was.

When it had gone as far up as it would go, she got up from her squat, turned around, and bent over like a defensive lineman. At that point, her hands and feet were on the floor, her booty was facing us, and about 25% of the object was still visible between her cheeks. There wasn't a drum roll, but there should have been, because the next thing you know, her handler walked over, slowly pulled the thing out of her; and her sphincter, attempting to regain its rightful shape, literally kind of went "ZZZZHOOOP!"

So, that was a little different from the days of working with Wendy and Bill …

AL ROKER, JR.

I was in charge of booking all of the XXX talent that was on the show or, every time we'd do a live broadcast or whatever, I was in charge of the porn stars. And, I knew a lot of porn stars. I had actually been going to the porno convention in Vegas for probably seven or eight years before I started with Mancow. So I was kind of challenging myself to find different things or whatever. So, at this time, we were on the 17th floor of the Merchandise Mart, in a small studio, then the producer's well, and then the band room on the other side. So I find this woman; her name is Abbey XXX. So, she comes in the studio, and we set her up in the band room. We've got the mics. So, I go in there. Turd goes in there; Freak goes in there. She's Abbey XXX, Queen of the Large Toys, that's ... So, we go into the studio, and she goes, "Anybody want a Coke?" And spreads her legs and pops a can of coke out of her vagina.

BRIAN "THE WHIPPING BOY" PARUCH

That woman actually "performed" onstage (albeit behind a screen, in silhouette) at the Allstate Arena at the Karnival of Kaos with Kid Rock, which is where I also Jello-wrestled some strippers, and later held up a gigantic, oversized poster of a male genital onstage.

Thank God, the invention of YouTube happened later.

J LOVE

I love the statement that if YouTube, or Facebook, or any of these social media entities had existed in the days of the Mancow's Morning Madhouse, like the one that I know, we would've been in jail and locked up for sure.

ROBERT CHASE

There was about a two-week period where I never saw (Mancow). I followed him, but I never saw him. I would be down the hall in the office—I would have read the *Tribune*, read the *Sun-Times*. I was ready to go, just waiting for the word that I could be let into the studio. He would slip out of the studio and go to his office with his staff. J Love or whomever would say, "We're done, last song, here you go." It was literally about a two-week period before we exchanged words, and the first words we exchanged were (with) him walking by as I was coming around to the door of the studio; he actually walked past the door, and I said, "Hey, fucker." Swear to God. He turned, without missing a beat, and said, "Robert Chase? Big fan." Maybe he didn't hear me. That was Mancow.

JEN JAMESON

Going from the laid-back Wendy and Bill and Whip showing up, like, five minutes before their show, to people rolling in at 4:00 and 4:30, and then Mancow storms in like five minutes before he goes on air, and all hell breaks loose … it was stressful.

CHRIS "PAYNE" MILLER, *Air personality*

Mancow called me. It was after the Rock had switched formats; I can't remember if I had just started law school, or if I was already in school. He called me, wanted to bring me over. The PD was Alex Luke. (Mancow) said he wanted to bring me over to do weekends, and to call Alex Luke. So I called Alex Luke. I thought it would be a great part time job. He was, like, "I don't really have anything for you." So I call Mancow back. I'm like, "He doesn't have anything for me." So, Cow's like, "They do 'Worst of Mancow' from 6-10. We'll just drop that and you can take that slot. I'll call Alex." And that's how I got in.

Mancow pulled his show off so I could be there.

JEN JAMESON

There was a day that first summer where something went wrong with the delay, and I don't want to point fingers, but I think it was probably J Love who was doing all that crap at the time. Somehow, Mancow came in on one of his tirades. I happened to be exiting the studio at the time, and it all got pointed at me. He went on the air and something was jacked up; he had all his affiliates, and his first half hour was dedicated to making me cry. I went to the promo … area, because a lot of times I would sit and go on the computer and just sort of wake up before I had to drive home and whatever. I just sat there listening, and I wanted to quit. I wanted to die. It was horrible. And the next day, it was like he's your best friend again.

MIDGE "DJ LUVCHEEZ" RIPOLI, *Technical Producer (Mancow's Morning Madhouse)*

We were up on (the 17th floor), and we had two wrestlers in … they were a tag team thing. And we had a bunch of strippers in from some club. We put the strippers in the green room.

The wrestlers came there by themselves, so I'm like, "If you guys want to go down to the conference room, there are some girls down there. They're all just waiting down there — we'll come down and get you in a minute." So, we took the guys down there, and Cow's like, "Hey, go get (them)." I go running down there, open up the door of the conference room, and one stripper's lying on the table, completely naked, and they're doing lines off of her chest. Swear to God. And it was like, oh, you guys have a little party in here, huh? And they did the lines off her chest, and went in and did the show. That was probably the craziest thing I think I've ever seen. Those guys knew how to party.

JIM "JESUS" LYNAM

That's a completely true story, and I don't think it was cocaine. I believe it was crystal meth. It may have been the one and only time in my life that I've seen crystal meth. It was funny—I was watching a show the other night. One of my friends goes, "I've never even seen crystal meth," and I go, "I'm pretty sure I've only seen at once … at work." If you think about it, how many things can you say after being in that atmosphere, "Wow, I saw that at work?" I didn't see it in Tijuana; I didn't see it in some dark alley. I saw it at work. I have friends, (who) talk about being in a frat and how crazy it was, and I didn't go away to school –I was in Columbia College. I always tell them, "I didn't have a frat—I hung out with guys named Turd, and Freak, and Mancow, and nothing you're going to tell me is going to top any story that I have." It's not being jerky; it was just that insane.

FREAK

Roker did sports. I did the music and the traffic. Brian did current events kind of stuff. We all had our own set jobs, so that there was really no overlap. Because I was a music junkie, I was out 5-6 nights a week. I would sleep in the afternoon, and be out all night. So I would be out at the Double Door, scouting new bands, Metro, House of Blues. I'd go see Hall & Oates, Pat Benatar; it didn't matter. I would just go to a show just to go see it. You never knew who you were going to trip over—stars or whoever. You don't find them in your living room; you've gotta go out.

While dozens of characters and personalities appeared on Mancow's show throughout its time on Q101, none reached a level of popularity like Jeff Renzetti, who more famously went by the on-air sobriquet of "Turd."

AL ROKER, JR.

Turd is…a larger than life character. And (he lives) up to that character. So, it's accepting every shot at the bar that's given to him; it's flower pot on the head, running around in your underwear. It's living up to that larger than life character. Jeff Renzetti is the nicest guy you'll ever meet. That's the guy who you want in the foxhole with you.

BRIAN "THE WHIPPING BOY" PARUCH

He would do anything; he was afraid to do nothing. He would eat more than any human normally would be able to eat, drink more than any human would be able to drink; he always had a cigarette in his mouth, and he was actually very nimble for a big man. He was beyond belief, very, very, funny, and witty, too.

J LOVE

Natural comedic talent and I think he had something that Mancow was able to exploit and bring out the best. Turd has like this comedic timing and the ability to play the dumb guy, to play it straight and convincingly. Really, he's just a big teddy bear, like, warm-hearted, fun loving …

BRIAN "THE WHIPPING BOY" PARUCH

In real life, he was a lot smarter than just a dopey guy who did stunts. He was actually a pretty smart guy who had a very good sense of humor. A cynical streak to him, for sure, but he knew that his living was in doing stunts, and therefore, did them.

FREAK

I knew Turd before the show, and he was a lot like me. I would say that him and I, our characters are most close to our actual selves; what you see is what you get. I wouldn't speak for anybody else, but for me, a lot of what I said on the show, I didn't have to think.

JIM "JESUS" LYNAM

Maybe the funniest human being on the planet, I think he should've been on *Saturday Night Live*. He had comedic timing down to a tee; he played his role better than anybody I've ever seen. You would never know that Turd was one of the most well read people on our show. He took every book home and read it. He was that blue collar, south-side guy; that made people love him. I think he was such a connection for people out on the street; he was that guy people could touch.

He was so good at talking to people, and then when he was in the studio, he never missed a beat, ever. You threw a joke at him—he knew exactly what to say and when to say it.

BRIAN "THE WHIPPING BOY" PARUCH

(Mancow) did create his little professional wrestling world saga thing with all of us (as) characters and stuff, which was, to tell you the truth, a pretty ingenious thing. It really was. At the very least, it was unique.

AL ROKER, JR.

I think we all did get along. I think it was—I mean, there was Mancow, and then there were the rest of us. And I think the rest of us were united in the fact that we knew that we were the rest of us.

BRIAN "THE WHIPPING BOY" PARUCH

I was never one of them. There was a lot of fiction to that, but it was also true. I had not been at Rock 103.5. I didn't care about strip clubs; I didn't want to get drunk; I didn't want to do all the things they did. In the back of my mind, I never wanted this show, even though I did appreciate all it did for me.

CHRISTIAN "CAP" PEDERSEN

Another thing that listeners didn't know about Brian was that he would always get booked to do all sorts of beer and liquor appearances at bars and clubs throughout the city. The big secret was that Brian didn't drink. In all the years I knew Brian, I don't think I ever saw him have one sip of beer. There are not too many people in FM radio that don't drink.

MIKE ENGLEBRECHT

There was a buzz when we would show up at a bar with one of these guys to do something. We all kind of benefitted from the glow around the station and the people we were working with.

JEFF DELGADO

After a while, things turned into a party non-stop, both in and out of the office. This was especially true when we Promotions Assistants were assigned to do bar nights with Turd and Freak. When we would be scheduled to do a two-hour appearance, we knew it was going to be at least a six-hour appearance, maybe even more, because it involved heavy drinking, chicks and drugs. Drugs weren't offered all the time, but you could count on beer distributor guys offering tons of free beer and liquor along with groupies wanting to drink with us and even hook up.

ALEX LUKE

Mancow helped us focus on a younger male end, and get a cohesive sound to the radio station; (he) helped me justify my desire to go there, maybe that's the better way to say it. We didn't have the music just prior to my arrival, or even in the first year I was there. I felt like we were searching a little bit for our position in the market. In hindsight, I wish that we had gotten there quicker.

MIKE ENGLEBRECHT

Alex Luke was a great guy, but he was dead the minute (Mancow) showed up.

PHIL "TWITCH" GROSCH

Anyone for this book who tells you that they could (manage Mancow), or did, is flat-out lying… To some extent, Dave Richards could manage Mancow. But nobody else could.

ALEX LUKE

From my perspective, he was … I will say with certainty I didn't manage him. I was part of the team that attempted to manage him. Part of the reason he was so great was because he was the extreme—he was very real, you didn't know what you were going to get, which I think was an actual positive, and he super-served the people that tuned in for him every day. He knew exactly what they wanted, and he gave it to them on a regular basis. It's one of those basic rules of radio: You identify your audience and you give them exactly what they're looking for every time they tune in. And he did that.

FREAK

(Alex) didn't last too long, just because that whole morning experience was a little more than your average human could take care of. For all the great things we did, there were always people complaining. Dealing with all that, that just takes a special person. I wouldn't want to do it; I'll tell you that.

JIM "JESUS" LYNAM
The amount (Q101) had changed just by the fact they hired Mancow was amazing. I was hired by a guy that I actually liked, a guy named Alex Luke, who I found to be a real genuine guy, (who) knew a lot about music, but I don't think he really knew—he didn't hire Mancow, so—I don't think he really knew how to handle that kind of morning show atmosphere.

MIDGE "DJ LUVCHEEZ" RIPOLI, *Technical Producer (Mancow's Morning Madhouse)*
Cow liked Alex Luke, I think.

ALAN SIMKOWSKI
Alex Luke lived happily ever after.

Alex Luke did, in fact, live happily ever after, jumping from Q101 to a handful of enviable positions, including Vice President of Music Programming at Napster, Director of Music Programming and Label Relations for iTunes, and Executive Vice President of A&R at EMI Music. If there were a contest for "most successful post-Q101 career," Alex Luke would easily land in the Top 5, and possibly end up running away with the whole thing.

Not long after Mancow left Rock 103.5 to join Q101, Rock 103.5 changed format to the non-descript, born-to-fail, "Jammin' Oldies" format. Chris Payne was the first to make the jump from 103.5 to Q101. Left on the beach, but not for long, were Rock 103.5 DJs Sludge and Ned Spindle, as well as the station's tenacious, no-bullshit, alarmingly cool-under-pressure Program Director, Dave Richards.

Richards replaced Alex Luke in a transition so quick and seamless, it was apparent that his hire was in the works long before Luke was "wished well in his next endeavors."

DAVE RICHARDS, *Program Director*
(My goal walking in to Q101 was) to create a vision, and that was to own 18-34 and 18-49-year-olds in Chicago, both men and adults. It was also to try and mesh this eight-year-old radio station's music heritage of alternative and this hard rock, "black hat" morning show, try to make them one. That wasn't going to be easy. Very, very, different cultures. Mancow and I came from a very different culture

than Q101 was. So it was bringing cultures together, and bringing a cohesive radio station together. All sides were going to have to give in a little bit. Q101 music wasn't necessarily going to have the same sort of female appeal. It couldn't be two separate vessels.

JIM "JESUS" LYNAM, *Producer (Mancow's Morning Madhouse)*
Dave Richards was the best, just because he had already dealt with (Mancow). I think, when you're a new Program Director, and you come in, and you're coming in from the outside, your natural reaction is, "Pfft, I've got it. I'll show this guy what's what." Then you get there and you realize, "Wow. I don't know if there's any way to do that." Dave had (Mancow's) trust.

ROBERT CHASE, *Air personality*
Dave was great in that I knew what Dave wanted, because he told me. With Bill Gamble, it was more like, he hired me, and he expects me to know, so just go do it. With Dave, in early conversations, he told me, "As long as I'm here, you're going to be here." When your boss tells you that, you're hoping he's going to be there a long time. There was a quota for certain things—he wanted phone activity, interaction with listeners. We'd meet at 9:30 every morning, a half hour before going on. He wanted to know what I had planned, so I had to have something planned: phone call wise, generate some topics, think about things. You got feedback from Dave. Dave was a guy who would leave a note for you, saying, "That was great—great bit." You knew with Dave when you did something well. When something needed work, he wouldn't just blindside you with, "That sucked." He'd say, "You could've tried this, or you could've tried that." With Dave, he wanted you to try. He wanted you to think outside the box and (he) wanted you to try doing things you maybe hadn't thought of before. It lifted my energy, because with Mancow, came Dave Richards. Really. So, following Mancow and having Dave in the driver's seat as a boss, really giving me some guidelines to work with, I had to elevate my game and maybe be a little less laid back—not match Mancow's energy, but bring it up a little so that the energy generated by the morning show would be carried through the day.

ABE KANAN, *Imaging/Production (Mancow's Morning Madhouse)*
Anyone who's a GM or PD, for the most part, is a failed radio guy. They just want to put their stamp on it. That's why Dave Richards was so good; he hired

people and let them go, let them do their thing. He wasn't afraid to give them that power.

JEN JAMESON, *Air personality*

Dave Richards helped give me my name. In college, I was Jen Longawa; it wasn't unusual to be just yourself in college. And then when I went to Fort Wayne and Omaha, I was Jen X, which in 1997-98, was cool. And Alex hired me and was, like, "Well, Jen X won't last. I think it's already getting a little dated." He said, "Why don't you just be Jen? We'll figure something out as we go." It was never really a top priority, so I was just Jen. That was it … So then, Dave came in, and that was one of his first things, like, we need to give you a name. This was a process that took weeks. I literally went through phone books and the old big radio guides that had everyone listed from the stations… I went through looking for ideas, like the typical generic radio names that you could put with any first name. I had to keep Jen. I couldn't all of a sudden become this other identity, like Electra. I narrowed the list down to maybe ten names that went well with Jen. Dave narrowed it down to Jen Jameson or Jen Ventura. But, he wanted me to be Jen "The Body" Ventura. I said, "Dave, if I'm going to be Jen 'the Body' Ventura, I can never get rid of that." And I hated wrestling; I always despised the wrestling culture. So that was the last thing I wanted to be. So I said I'd rather do Jen Jameson than Jen Ventura. He said, "You'd rather be associated with a porn star than a wrestler?" I said, "Yes; any day." So that's how I became Jen Jameson.

MIKE BRATTON, *Imaging Director/Production Director*

Dave came over and brought with him a few people from Rock 103.5, so when he came in the door, it was me doing imaging and Art Wallis doing commercial production for the station. When Dave came in, he basically had me re-audition for my job, which was a little distressing, to say the least. I knew full well, when Rock 103.5 went away, Ned Spindle was there; I didn't have to be a genius to do the math that more than likely, at some point in time or another, Dave was going to rescue Ned and bring him over to the station. What I didn't know was that he was going to ask me if I wanted the Commercial Production Director's job before he told the Commercial Production Director that he was out of a job. So I got to sit on that knowledge for 24 hours ahead of time. I had to make a decision right then and there. I had to say, yes, I want that job and I want to keep being gainfully employed, or no, I don't want that job and definitely be losing my job right then and there. It was a much friendlier conversation than that, but that was the gist of it. So, my decision was relatively pained, but it was really about the only thing I saw I could do. … That was a total shit-show …

NED SPINDLE, *Imaging Director*

(Q101) was not the same as Rock 103.5. I'm sure it was more like Rock 103.5 than it had been a few months earlier, because Dave Richards was there; so, I was stepping into a job with the same PD as I had at my previous job.

CHRIS "PAYNE" MILLER, *Air personality*

(Q101) was really different from what I was used to at the Rock. The two stations were really very different. Rock 103.5, bunch of metalheads. We were playing a lot heavier music. And Q101 back then was truly alternative … there was no rock on it.

J LOVE, *Producer*

I would say there was a pretty distinct separation feeling of church and state at that point. Depending on your perception of which was stronger, the big burly morning show, the monstrosity that was Mancow's Morning Madhouse, or the rest of the station, it felt like, at 10:01 or whenever we chose to get off the air, then the rest of Q101 would kick in. The perception went from a family to a friends and family situation; there was now two distinct groups operating within the same four walls. I was fortunate enough to have a foot in both groups. I did the Mancow thing that was the lions' share of my duties and involvement, and I was very engrossed in that culture. I was also an on-air personality where I flew under the radar of J Love, and essentially, I had the opportunity to be kind of in both the family and friends camps at that point.

KEITH SGARIGLIA, *Promotions Manager/Webmaster*

I remember a lot of odd mornings where I came into my office, and there was a girl with no pants on, sitting in my chair, at my desk, because I was kind of over by their office. Or there was a woman topless with a snake wrapped around her. Things like that were kind of jarring at eight in the morning when you're getting to work.

JIM "JESUS" LYNAM

You could walk in one day and there'd be a chicken walking down the hall, and you'd go, "Why is there a chicken? Oh ... Mancow's show. Why are there ten clowns and a Mexican rap group called Los Marijuanos? Well, it's Mancow's show."

FREAK, *Air personality*

There'd be strippers running around, wild animals in the hallway, or something going on.

MIKE BRATTON

(There were) a few times when I walked in to strippers changing in my office, and not good ones, not strippers that I would be, like, "Nice, you're changing in my office." It was like, "Aw, come on. I'm going to have to fucking fumigate this place now." Other than weird moments like that and dragging me in to watch gaping asshole lady, they were decent enough to me. They weren't overly jerky to me directly, but the overall feel they brought to the rest of the place was pretty icky.

MIKE ENGLEBRECHT, *Promotions Assistant*

There were naked women everywhere. When I say naked women, I don't mean just topless. There were days where you'd walk in and there'd be five or six women completely naked head to toe, standing in the hallway, waiting to be called in to the studio.

CHRISTIAN "CAP" PEDERSEN, *Promotions Assistant*

A male listener made oral love to himself in the Q101 lobby during office hours.

JIM "JESUS" LYNAM

We had an alligator in there one day; we had monkeys, we had every stripper known to man, porn stars, Roker (Jr.) got peed on ...

CHRISTIAN "CAP" PEDERSEN

So many fantastic memories, but as a collective, it was all the naked people that willingly exposed themselves to me (that stands out).

MIKE ENGLEBRECHT

It wasn't just strippers; it seemed like the star power ratcheted up. I remember one morning walking in and there was Bret Michaels strumming his guitar, softly singing "Every Rose Has Its Thorn," kinda warming up to go on Mancow's show, right there in the foyer … I kind of get chills, thinking about it today.

PHIL "TWITCH" GROSCH, *Promotions Assistant*

The station was never the same after that, for good or worse. Everything that anyone who listened heard on the air, it was exactly that, and even more off the air. I mean no exaggeration: naked girls in the hallways, drinking and intoxicated people, whether part of the show themselves or entourage people. At any hour of the day, people could be drunk and intoxicated and doing who knows what else behind closed doors.

NED SPINDLE

Mancow, for some reason, in all the years that I worked with him at two different radio stations, never gave me any shit. He was always really nice to me. We weren't close, and we never hung out, but every now and then, once every month or two, he'd walk into my studio, lean against the board or the wall, talk at me very emotionally about something, and then just walk out.

DAVE RICHARDS

(What do I remember?)

A naked guy on stilts holding a chicken standing in my office stays in my mind. Two female porn stars being thrown into my office while I was on the phone, and I couldn't get off the phone, and (Mancow) just encouraging them to "get it on," and then closing the door behind him. A naked Al Roker (Jr.) lying down on my desk while I was on a phone call …

I don't know that there's ever (been) one specific thing.

ROBERT CHASE, *Air personality*

There were often times when Mancow's Morning Madhouse was better heard than seen. To have close proximity to some of the goings-on was just not a pretty thing at 10 o'clock at night, let alone 10 o'clock in the morning.

MIKE ENGLEBRECHT

It was a lot of fun, still; it was just fun in a less clean kind of way. I was in a role where, as a P.A. (Promotions Assistant), I'd be driving Turd around. I was one of two or three guys that would have to get there—we had a rotation, and when it was your turn, you'd have to get there about five in the morning and sit and wait until Cow decided where he was going to send Turd, then hop in the van and take him out.

ALAN SIMKOWSKI, *Director of Market Development*

There was a period of adjustment, but it is what it is, you know? Unfortunately, I think (Mancow) leaving the station (in 2006) is what ultimately took the place down, because they really thought they could succeed without him, and they didn't. I would've found a way to make that work.

MIKE BRATTON

It was okay to be just a complete dickhead to anyone that you wanted, just because, "I'm part of this show, and we're the best thing that happened, and we're going to save this radio station," which it didn't. It just really fucked up the culture of the place, I think. It changed the music of the place, too. Then it became, well, "Depeche Mode doesn't really fit with the Mancow listener." "Well, who gives a fuck?" Or, "Soul Coughing doesn't really fit with the Mancow listener." "You're right, but it does fit with the people who are here for the other 20 hours of the day." It started pointing Q101 in the eventual direction of thinking that adding a Nickelback record was a good idea, and calling it alternative because that's what the station was still listed as in the trades. It really just completely warped the place, and then, it just warped my perception of the station entirely, and I feel like it really confused the listener, too. It confused the overall marketplace. We weren't helping by doing liners like, "Mancow in the Morning, Q101 music all day," or some bullshit like that. What did that even mean? It became increasingly difficult; you could actually hear the disconnect in the culture at the radio station behind the scenes on the air, throughout the rest of the day.

DAVE RICHARDS

You had such a big, giant voice in the morning that you were going to piss off some people. I'm sure the day Cow landed on Q101, there were a few people who said, "Nope, I'm gone, I don't want this." But then again, he was bringing over the biggest male audience at the time in Chicago. It was going to bleed into the rest of the day … (The audience) came along for the ride, and I think they came along in the same fraternal order with the same smile on their faces, the same inside jokes: yeah, we're part of this; we get the sense of humor, we get the attitude. And they came along for the ride.

In August 2011, I was interviewed on ABC television's Windy City Live *about Q101's demise and legacy. Towards the end of the interview, host Ryan Chiaverini asked for my opinion of Mancow. I felt then, as I always have, that the culture around Mancow completely distracts from the simple fact that Mancow is a gifted, results-getting personality. As I told Chiaverini on the show, "Mancow gets pegged as this radio boogeyman. In truth, Mancow is, hands-down, one of the most talented people I've ever worked with."*[1]

JIM "JESUS" LYNAM

There were a few times in Q101's history, I think a lot when we were on 17, and a couple of years down on floor two, when you had a group of people, and no matter what the problem was, (Q101) was a machine. It rolled. Mary Shuminas as Music Director, James VanOsdol, Robert Chase, the sales people got it … There's that feeling where you come to work, and you're a 28-year-old kid, and you're in awe of how everything moves, and everything goes. I don't know how many people have that in life.

TIM VIRGIN

The station was now all about what (Mancow) was, and not about what I loved and wanted to work for at Q101, and I was sad about it, you know. Without getting into details, I was stupid, and talked shit.

Dave Richards walked up to me right when they hired him … he said to me in an elevator one time, "I like what you do. Don't get freaked out. I'm not the bad guy yet." He was amazing. If anybody could ever handle Mancow, it wasn't going to be Alex; it was going to be Dave Richards, and that was the reason why Dave Richards was there.

When my contract ended, that was it.

---

[1] Windy City Live, "Time Out Chicago: Alt Rocked," 16 August 2011.

Tim Virgin's Q101 story would resume over a decade later, when he returned to the station in 2009. His replacement in Afternoon Drive was one of Dave Richards' most trusted and dependable air personalities at Rock 103.5: Sludge.

SLUDGE, *Air personality*
(Coming to Q101) was very weird at first. Like, "Oh my god, this is like maybe going into Iraq and running Iraq." Everybody at Q101 was awesome, especially JVO. It was definitely a tense feeling, though. The first months were like, "These A-holes are over here now?" Q101 certainly felt like they won the war, but a lot of us came over there.

DAVE RICHARDS
Sludge is … sports-meets-music-meets-the-goofiest-sense-of–humor, and the most likeable human being in the world.

ROBERT CHASE
Sludge was fun, another good guy that brought a lot of energy to afternoons, very dedicated to his craft. (He) had a unique personality, and (was) a fun guy to be with.
(He's) a guy who, to this day, has a lot of radio credibility in that he works hard. He shows up and he puts in the effort before the show, during the show, and after the show.

ABE KANAN
He's the guy that I learned the most from in radio … he's the nicest guy you'll ever meet in your life. He was one of my first supporters; he loved everything I did.

MIKE BRATTON
Sludge was a fucking goof, and somehow or another, he had game; like, he pulled in a lot of really, really interesting ladies.

RYAN MANNO, *Air personality*
It's a hilarious subject on its own when you meet Sludge and you think like, "Wait—you're the guy that's talking about getting laid and summer dresses?" and he's this tall, dorky, awkward moron.

SLUDGE

I'm glad Ryan remembers me that way. I've always felt that I was certainly some of the "guy on the radio," maybe more so now; I'm all the guy on the radio, but just an exaggerated form. Using the name Sludge, the things I did, and the way I did my show was highly produced, with a lot of weird stuff going on. I always thought it was the exaggerated me. If there was something I liked, or something that I felt passionate about on the show, it became exaggerated in a lot of ways. It certainly wasn't always true, but the comments I made were the first things that came into my head.

ABE KANAN

To me, he should've been doing mornings on Q101. I think he's the best guy out there.

DAVE RICHARDS

I've never seen anyone who can create such a great environment around him for a lot of young people who want to get into radio and perform just like him, and he still does to this day.

MANCOW: IN HIS OWN WORDS

CHAPTER SIXTEEN

Q101's Program Directors and General Managers may have had figurehead job titles, but the reality was that, between 1998 and 2006, Mancow Muller ruled the station.

It wasn't just Mancow's ratings that put him in the driver's seat; it was also his notoriety and brand awareness in the market. Through word of mouth, marketing, and a fearless media persona, Muller's star rose fast and dramatically enough to gain him membership in Chicago radio's Red Carpet Club (an elite—if metaphorical—group that also includes fellow radio rock stars Steve Dahl, Jonathan Brandmeier, Eddie & Jobo, Roe Conn, and Kevin Matthews).

Beyond that, once you consider that Mancow's salary dwarfed the combined salaries of Q101's senior management, it's not hard to figure out who was really behind the wheel.

Despite his successes at Q101, Muller bristles at the thought of looking back on his time at the station. For every victory achieved, there was also struggle: personality clashes with management, internal clashes with morning show personnel, ongoing culture clashes inside the station, FCC conflicts, witch-hunt level scrutiny after Janet Jackson's nipple was flashed on TV. At some point during Mancow's time at Q101, radio stopped being fun.

Mancow made it clear, once this book's premise was announced, that he wasn't interested in reliving, or reminiscing about, his Q101 years. Even in casual conversation, he frequently steers the topic of Q101 back to the present day and what his future plans are. As he told me, "Talking about Q at this point feels like I'm trapped next to some fatass reliving his glory days playing high school football. Tedious."

After some back and forth (and through gritted teeth), Mancow agreed to be part of the project in the interest of "setting the record straight." There were some parameters around his involvement, though: most importantly, he didn't want to be interviewed directly for the book. Instead, he agreed to forward me a handful of thoughts and observations about his time at the station.

Those thoughts came to me in furious SMS bursts, which I went on to format into something more document-like.

Because his quotes didn't fall directly in line with the questions and narrative I was building for the book—and because he's Mancow—I decided to isolate his thoughts in a chapter all his own.

Mancow interviews Mancow ... for real.
Feelings when your contract ended?
Contracts end. All contracts have an end date.

You knew the contract would end ... the surprise was that you believed they were fellow artists and friends, many of them. Then, before you were out of the building, they really went after you as an enemy.
That was the shock. For a year or more, they planned how to neutralize me. Their surprise was how, without their help, and in many cases (with them) working against the show, the show's ratings continued to grow *huge*!

You didn't care about leaving, just how ugly "the suits" got. How did you feel?
It wasn't until later really that the betrayal sunk in.

In reviewing people - you only spoke about a few people.
Silence on the rest can speak for itself.

You shouldn't admit you've forgotten almost everybody else; that might seem shallow. You aren't shallow, just too busy with life and kids to look back. Intriguing, that you still now feel pain?
I would have thought it was all over now, baby blue.

You shouldn't have given into this. Why let them talk to you for a stupid book? You wouldn't read a book like that! Wait! Would you?

The state of radio in 2012
Radio has become so diluted and boring. To work in radio now is to work in a factory.

Being the best at radio in 2012 is like being the best buggy whip maker in 1896 at the dawn of the horseless carriage.

I was watching *The Bride of Frankenstein*, and I was taken by how that movie is so symbolic of (what) all radio has become. These corporations gobbled up all the radio stations they could. Then they cobbled them all together. Now this corporate radio-monster stumbles around, unknowingly killing everything in its path. Frankenstein was not inherently evil; just stupid, and people got hurt.

On Q101's demise, and the nostalgia that followed

I had several offers on several occasions to do "Goodbye to Q" shows. Why? Self-indulgence? I also believe it's not over yet. How about that? Just you wait and see! I have concrete evidence that it ain't over yet.

I strongly dislike talking about Q or the people involved. I don't live my life in the rearview mirror. My best show is yet to be done.

Mancow's Memorable Music Moments

Death Cab for Cutie didn't want anyone to look at them. We had to pull the drapes so they couldn't be gazed upon. The band's lead singer, Ben, had a side project called the Postal Service; that is one of the best CDs of the last 20 years (along with Joseph Arthur, Ike Reilly, Tally Hall, and Sigur Ros).

Well, the other members of Death Cab *hate* Postal Service! Ben pleads with me not to talk about it. He was kind to me and in a tough spot. "Are they jealous?" I inquired. "Uh huh." Then I saw them live in performance. I introduced them. Awful group. I witnessed Death Cab droning on, looking at their feet. The audience was too cool to act interested, so they stared at their own feet ... or at anything *besides* the band. That's an alternative crowd for you! Boring hipsters so lost in their own irony that they can't actually acknowledge their own likes.

Kid Rock would not exist without my show. Period. I'm very proud of that! The record company called and ordered me *not* to play ("I Am the Bullgod") because they "were no longer working the project." They were dropping him. I didn't stop! By the time I played "Bullgod" and "Cowboy," they were old forgotten songs never heard on the radioscape. The listeners loved him. He still treats me like a saint, and is beyond kind to me.

On not socializing more inside the station

Why didn't I hang out more? I was *working*! It took great effort to create hours of the best radio in ChiCOWgo history. Many hours of work went into every hour that the show was actually on-air. It wasn't like *Full House*, where Uncle Jesse would click on the mic and hilarity would ensue.

On his critics

Wanna play radio? Grab a cell phone, website with news headlines, a CD player and go in your closet. Now YOU be clever for 25 hours! That's what I do every week. I feel like the Eloi when I leave my studio.

Radio management

Middle management (people) in radio are failures usually. They failed on-air, and now they are going to tell *you* how it's done. They are mostly envious types and can't fathom the relationship and almost supernatural bond I have with my listeners. It's like getting to the panty and always having to stop. They are frustrated and confused. They couldn't make it in a studio, so now they are a placeholder in some cubicle. These guys failed, and now hate those of us that can make sweet audio love to the masses. Mancow comes through your ear and settles in your soul, to some a hero, to others a heel. It makes no difference to me! I get people to *feel* something, and in this desensitized world, that's really something.

Mancow's Morning Madhouse Members

Freak? I gave him that nickname. Nicknames on air allow performers to step out of themselves. He was limited in range, but attached to me, he had some value. Bill Gamble (he was a competing suit from another station) called me after he paid Freak to go on *his* station to bash me, and asked, "Did I get ya?" I laughed and I said, "Oh, yeah! You got me! But let me ask you Bill, do you respect a guy that would do what he did to me?" He said solemnly, "No." I called my "friend" Freak, that I had found under the rock of obscurity, and asked him, "Why?" His answer was, "Awww man, it's just business." What a bastard! This is from a guy who lied telling us all he had severe back pain that prevented him from working with us! I let him out of his contract so he could do that?

Turd? I gave him that nickname for his foul stench. He was a greedy, nasty, shallow, slovenly ignoramus. (You know, like most public school-educated southsiders.)

I wrote most of what he uttered on the radio via little scraps of paper. (He was) a drunkard that was always late and (who) always had to leave early. A barback when I met him, his ego soon ballooned to nasty proportions.

A superstar in his own mind! Ants have the ability to pick the sugar from the sand and I will attempt to do that with him ... he was a savant when it came to timing. *Nobody* on Earth knows how to time a punch line like that slob. He never thanked me for plucking him from obscurity. Not once. Sadly, the last I heard he was working a trash truck. It's an honest living, but had he not been so quick to stab me in the back, he'd still be a radio star.

Luvcheez (a nickname I gave him) was a male stripper in San Freako, Ca. when I found him. A man of diminutive stature, we used to bill him as "Spunky the Exotic Erotic Midget." Attitude equals altitude, and he had the worst. Given to "little man syndrome" and violent eruptions of temper, he was disliked by many. I, however, think he is a genius on a mixing board, producing Mancow phone scams, parody spots, and songs, all masterfully. He was George Martin to my Beatles, Radar to my Henry Blake, sour cream to my chive. He had superstar talent but constantly gave in to his inner anger and base side. Sad, because I cared for him like a brother. He had a magical skill of anticipating my every move ... except when I fired him. It eventually imploded after one of his emotional outbursts. I actually used to pay him an attitude bonus if he was pleasant to me! "Just say good morning and not grunt." Hilarious! Him, I truly miss for some unknown reason. Ha!

Jim Jesus (a nickname I gave him) was void of humanity. I hope he finds God. Really. I'd like to take him to church.

Al Roker Jr. (the nickname I gave him) was a rotund loyal fellow of enormous loyalty. Love him still! He also booked many guests for the show, like a braying mule. He bellyached often but was wonderful. Yeah! Yeahhhh!

Q101 co-workers

Robert Chase and Chris Payne were something I never was: *cool* … I'd say Chase and Payne always made the listener want to fire up a blunt and hang with them.

Who is Mancow?

When I relinquish my body, my legacy won't be my stupid radio show, but my children. I sold off my possessions years ago. Flesh is only a cage that holds our spirit. Mancow is an Andy Kaufmanesque creation. Mancow is ego and bluster, the Hulk to my Bruce Banner. The real me has successfully shed most ego and fear. When I shake this frame, as we all must, I hope to blossom like a lotus into pure spirit and Christ-consciousness.

Never could I grasp people's occasional anger at the Mancow. (Most emotion directed towards me is pure love.) As much as I delight in eliciting any emotion in this unfeeling age, anger still amazes me. You take someone called "MANCOW" seriously? Hilarious!

Throwing a rock at the lunar reflection on a pond isn't the same as hitting the moon just as listening to Mancow is far from actually ever knowing me.

JAMBOREE '99

CHAPTER SEVENTEEN

Q101 put on its most memorable Jamboree in 1999, for reasons it never could have planned.

Red Hot Chili Peppers, Blink-182, Hole, Blondie, and the Offspring were among the A-listers on the bill that year.

JEN JAMESON, *Air personality*

I peed next to Debbie Harry at Jamboree. I had the mic, and they were doing the remotes to all the jocks hanging out around the venue. All of a sudden, it was, like, "Hey, Jen, we're going to go to you." I was, like, "Yeah, I'm in the bathroom," and Debbie Harry was in the stall next to me. I just sort of … "Hey, what's up, Debbie?" and put her on the radio in the bathroom.

DAVE RICHARDS, *Program Director*

Earlier in the day, Courtney Love catches Deborah Harry in her eyes for the first time. If you've ever spoken to Courtney Love, that gaze in her eyes -- it's really hard to describe what those eyes look like. She sees Deborah Harry, and she walks up and drops to her knees, buries her face in the crotch of rock royalty, and says something to the effect of "I pray to your alter," which puts the other 25 people who were in the area in a very weird place.

ROBERT CHASE, *Air personality*

Courtney had said (from the Jamboree stage), "You're going to love the Offspring coming up next, because they have so many great songs," obviously, facetiously delivering that line to a crowd (where) some maybe got it, (and for) some, maybe it just went right over them.

DAVE RICHARDS

I can remember two very clear things: Offspring being Offspring, and saying, "Hey everybody, throw your trash on stage," and then there being an incredibly long period of cleaning, and then mopping, and then waiting for the mopping to dry, and people still throwing shit. And JAM Productions asked us very nicely, "Why don't you get Mancow to go out on stage and ask people to stop." I saw something similar to a mob scene in *Animal House*, of it just getting worse.

ROBERT CHASE

When the Offspring noticed that there was a lot of garbage around, (they) ended up going on stage and ending their set by asking the crowd to clean the place up, pick up all the garbage around them, and throw it up on stage. And it came for what, 20 minutes solid, if not longer. It was just unbelievable that that much garbage was actually available.

MIKE ENGLEBRECHT, *Promotions Assistant*

Dexter says something to the effect of, "Throw all your garbage up on to the stage," which at first was kind of funny. From where I was sitting, the way the theater sloped from low to high, you could kind of see this … for lack of a better way to put it, it almost looked like a tsunami wave of garbage coming, and it was slowly getting closer and closer.

DAVE RICHARDS

I've never seen a barrage of garbage like that in my life.

SLUDGE, *Air personality*

When Offspring did that, Offspring was the second-to-last band; the Chili Peppers were the headliners.

In the Offspring's last song, Dexter told everybody to start throwing garbage up on the stage—at them, even—"whatever you have, start throwing garbage up here." (There were) 30,000 people throwing tons of everything around them, on to the stage.

I actually remember lying on a couch in the side green room. There were no other Q101 DJs around; I was alone in the room. Somebody ran in from the Tweeter Center and was, like, "Somebody's gotta get out there—you've gotta stop this, or the show's going to end right now." I was kind of dazed; it was one of those Jamboree days, which were like 15-20-hour days. I was, like, "What are you talking about?" I looked out the window to the stage, and all I saw was garbage being thrown up—I didn't know what was going on exactly. So I ran down to the side of the stage with this guy. The Offspring got done and walked off, and the garbage kept coming. He went, "You've gotta go out there and tell them to stop."

MIKE ENGLEBRECHT
As we were scrambling to get out of there, I remember Sludge coming on stage. I'm not sure exactly what he said, but I remember it didn't get the exact result I think he was hoping for. I think it just encouraged people to throw more things, to become more frantic and crazed, and to behave like idiots.

SLUDGE
I don't know why I didn't fear it at the time, so I walked right out to the mic, into the garbage rain, and said, "We appreciate your enthusiasm." I should have stopped right there and walked off, looking back on it, because maybe the garbage would have stopped then. Instead, my next statement was, "If you don't stop throwing garbage, the Chili Peppers won't be able to come on." A new amount of garbage tonnage came at me. That was just like daring people to throw more, and that's exactly what everybody did. It went on for another ten minutes, (and I) got hit with garbage, objects, and water bottles. Somebody even got a garbage can on to the stage; a full metal garbage can rolled up and made it up to the stage. I laughed about it, of course, to try and bond with (the fans), but that was the dumbest thing ever to say after "We appreciate your enthusiasm."

DAVE RICHARDS
Sludge ran out there to try and get them to stop, and he got exactly what we thought Mancow would get—a whole bunch of sodas and so forth on his head.

ROBERT CHASE
The Chili Peppers had to come out, but the stage was trashed with full-blown sodas and burger meat—it was just unbelievable. I was standing right

next to Courtney and Melissa Auf der Mar and Courtney was pissed because she made some remarks, and I was, like, "It's a matter of one-up-man-ship now, huh, Courtney?" She was, like, "I'm so pissed." She'd thought she kind of owned the day in some way, and the Offspring just wiped that off the map.

THIS AIN'T A SCENE: LOCAL 101, THE PAYNE YEARS

CHAPTER EIGHTEEN

I resigned from Q101 in December 2000. As opposed to when I took over the show in 1995, there was a blatantly obvious person in-house to name as my successor.

CHRIS "PAYNE" MILLER, *Air personality*
I think I was shocked by the fact that James was leaving, and left to go to XRT … I think I was more shocked that he was leaving. I think that I didn't ask Dave (Richards) if I could take over the show until he was already out the door. It was after he gave his notice.

JAIME BLACK, *Producer*
I started out at Q101 barely 16 years old in 1998, and in 2001, Chris took over as host of *Local 101*. Though my title wasn't producer right away, I moved more and more into that position the longer I worked on the show. So, it's impossible to look back at the last decade of my life and not associate it heavily with *Local 101*.

Short version, I loved working on the show. I loved finding new bands to book. I loved breaking new music. I loved doing something exciting and fresh and different on market three commercial radio! It was such a cool experience; it's hard to summarize it succinctly in words. I think the most telling sign that I loved the job was that I'd start looking for bands to book for *Local 101* interviews weeks and months in advance. If it was January, I might already have talent lined up to come on the show through April. It was exciting work; I felt privileged to be involved.

CHRIS "PAYNE" MILLER, *Air personality*

It was hard. It was really, really, hard. It was hard following in James VanOsdol's footsteps initially because I would still get mail addressed to him. I would get mail addressed to him two or three years later. These fucking bands were too stupid to know that he didn't work there anymore.

I'd get a lot of "Well, JVO played us, why won't you play us?" I was, like, "I'm not into it;" so, that was difficult. Starting from scratch with music submissions—I think the first few shows that I was on the air I was playing the local music that I had left over from Rock 103.5 and Chicago Rock, and that was three years before that. I remember the shows weren't that great. Shortly after I took over, you know, bands don't care who's on the air—they just want their songs played. They didn't care about James; they didn't care about me; they didn't care about the girl who did it before him. They just wanted their music played. And I was okay with that, that's cool. They'll give you anything in the world to play their shit, so I got a lot of free weed over the years.

JAIME BLACK

Being host of *Local 101* was not like your typical on-air radio position. It wasn't something that any DJ could have done. You had to know the local musical community and climate. You had to know how to conduct an interview—or three or five interviews, in an hour and a half, and condense those into about 20 minutes worth of audio. But most of all, you had to know how to talk to bands. Chris knew all of that. He had a history of hosting local music specialty programming, having hosted the Chicago rock show at the late, great Rock 103.5. As such, some of his artist relationships transferred over to Q101, when you look at acts like Local H, Kill Hannah, and The Beer Nuts (of all bands). But more than transfer talent, we brought on a ton of new bands throughout the last decade of Local 101, and Chris was always down to do so. I sincerely believe he would've been bored to death with a typical four-hour format air shift. He needed a living, breathing, unpredictable program with a format that started from scratch every week, talking to Chicago's eccentric artist community. For this last decade of Local 101, Chris Payne was the right man for the job.

CHRIS "PAYNE" MILLER, *Air personality*

I think Rise Against, I can probably take 100% credit for that. They've so much as admitted that to me or brought it up on the air before. I think James was the first one to ever play Chevelle, even though they think it was me … they give me credit for that. Sorry, James.

JAIME BLACK

I absolutely remember (the) night, where Scott Lucas came up for a Local 101 interview and it turned into an impromptu private acoustic performance. I really don't think we requested Scott bring in his guitar that night. I've gotta believe we were just anticipating a straightforward interview. It's entirely possible I'm wrong about that, though.

Either way, neither Chris nor I were expecting a private acoustic performance that night. I'm sure there's a recording of the entire event on a CD or DAT or hard drive somewhere in Chris' or my possession. But it really was this insanely cool gesture on Scott's part, in that he absolutely didn't have to perform for us in any capacity, and he did. And like Chris recounted, he was playing requests! As if playing a private impromptu show in this tiny production studio wasn't cool enough, we got to make the set list. Looking back at the countless Local 101 interviews during our time on the show, that evening absolutely stands out.

CHRIS "PAYNE" MILLER, *Air personality*

I've interviewed a lot of artists over my career; before I got to Chicago I spent four years in Milwaukee, a couple years in Tennessee, a year in Florida, and four years plus in Mobile, Alabama, so I've interviewed everyone from Don Henley to the Cult to every band in the 80s : Firehouse, Warrant, Trixter, Alice in Chains, Soundgarden, Nirvana.

Scott Lucas. He came in. I don't know why, (but) I was scared shitless to interview Scott Lucas. I have no idea why. I liked his music; I thought it was really cool. I first met him with Herb from the Beer Nuts (who) brought him in on my very first interview. And he was so hard to interview; he was like the hardest guy I've ever interviewed, and it made me nervous. The fact that I felt nervous doing an interview made me more uncomfortable because I'm not used to feeling nervous doing an interview. And it was years before I could finally shake that.

I remember Scott called me and he's like, "Hey, man, can I come on the show?" He'd never called me to ask to be on; I was always hounding him. He was always very gracious, and glad to do it. And the conversation was always wonderful until we started recording, and then he got as weird as shit.

So this was maybe the first time he ever called me and asked to be on the show. He had something coming up; I think it was a local gig. I was, like, absolutely. So he came in, and he had an acoustic guitar. It was when I was doing the early interviews during the week; it was one of the production rooms, kind of small, not very comfortable. The mics weren't that great; they worked only about 50% of the time. So he came in and we did this really, like, incredible interview… where it was, like, it was, you wanted to cry. It wasn't sad stories; he was just being very honest and being heartfelt about his career.

I think I even took advantage talking about the earlier days in his career and he started playing some acoustic songs. He played another song and it was, like, "Well you want me to play another one?" We were, like, "Yeah." He was, like, "What do you want me to play? Just pick it." So we sat there with Scott Lucas playing acoustic guitar, singing his guts out, playing every song we requested … I remember looking at Jaime, who was the producer all those years, and he and I were both absolutely 100% amazed.

From that point forward, I became very comfortable with him, even in social environments, and we actually had a few more moments like that on the air and in the years to follow.

JAIME BLACK

Local 101 was a feature that fell in and out of fashion throughout Chris' and my time on the show. Sometimes the station or a particular programming staff was behind the show, and other times, we were basically on an island. That being said, *Local 101* was Q101's longest-running on-air program, and a heritage show, at that, having featured Carla Leonardo… James, and then Chris as hosts. So there was always a lot of history invoked when talking about *Local 101*, even casually, because the majority of serious Q101 staff knew that the show had a respectable past.

More than anything, *Local 101* gave Q101 credibility. As the station went through a decade of identity crises in the new millennium, *Local 101* was a constant, where listeners knew they were going to hear songs hand-picked by a staff that loved music, not decided by an out of state programming department or passing industry trends. *Local 101* could break new music just for the crazy reason that it was good, of all reasons! That was something that, sadly, the Q101 format at large made a conscious decision to stop doing at some point.

It was also a program that listeners and the station itself could legitimately point to as an actual launching point for bands. Looking back at the '90s *Local 101* era, it was clear that artists like Local H and Veruca Salt benefited from the program. In the 2000s, we were fortunate enough to be on the ground floor for acts like Fall Out Boy and Rise Against. I've got to think that might have helped the show's staggering longevity, knowing that there was an actual chance that a superstar act might very well receive early support and exposure on *Local 101*.

CHRIS "PAYNE" MILLER, *Air personality*

I was probably left alone 95% of the time. And the other 5% of the time that I was fucked with in some capacity, I think it might've been (Program Director, Mike) Stern. Everyone else left me alone. If it's not broken, why fix it? I was older than everyone else on staff; I'd been on-air longer than anyone else on staff. I was in law school or already a practicing lawyer by that time. I was just not someone that was going to fuck around; I was not going to say stupid shit on the air. I was going to do clean interviews. No one was going to use profanity. I was going to play the right music. I was not taking any chances. So everyone pretty much left me alone. I think because they didn't get what we were doing and they wouldn't listen on Sunday nights at 9 o'clock, no one even knew we existed, unless something went horribly wrong, like we were off the air, or our segment repeated, because, as you know, we didn't do a live show probably the last five, six, years of *Local 101*.

ANTHEM FOR THE YEAR 2000: THE NEW CLASS

CHAPTER NINETEEN

The turn of the century brought a fresh influx of full-time on-air talent, beginning with Pyke (who lasted all of a minute), and continuing with future mainstays Fook, Sherman, and Alex Quigley.

Upper management also went through significant changes in the early 00s, as long-time General Manager Chuck Hillier and Program Director Dave Richards left the station. As a curious footnote, Chuck Hillier was permanently replaced by another Chuck (Chuck Ducoty), and Dave Richards was replaced by another Richards (Tim).

ALEX QUIGLEY, *Air personality*

They made a change at night. They had just hired a guy named Pyke ... I guess I can tell this story, it's true. He moved here from out of town, and I remember being in Dave's office. I guess they hadn't signed the contract, technically, yet, but Dave (Richards) had already volunteered to pay his moving expenses. And then Pyke jumped across the street to KISS. So they paid for him to move, and then he jumped, after, like, six weeks. Dave was so mad. And he called me. I had just gotten, or was about to go, on the air, 6 p.m., down in Champaign, when booming Dave Richards voice (said), "Alex, I need you to do nights for at least the next two months. I can't guarantee you're going to get the job or anything like that. I just need you to give it your best shot. Fuckin' Pyke." And he went off on the story.

FOOK, *Air personality*

I actually wound up hanging with Pyke a few times in Chicago. And I liked the guy—he turned out to be a nice guy who just was not into doing radio. He was into being good looking and a nice guy. I think he eventually wound up … in Toledo, and has been very, very successful.

ALEX QUIGLEY

So yeah, I went immediately five days a week, that was late March, early April, maybe, of '01. And I was, like, '"Yeah, I'll do this." So, two days (and) I found a sub for myself at 'PGU. I'd already graduated, but I was still a program director. So, it was really weird—I don't remember much of that time in my life, (but) he liked me well enough to put me in overnights.

FOOK

I just sent in like a cassette aircheck to Dave Richards. A couple weeks after I sent it, he called me back. And then I sent him another one. Then, I think I sent him another one. And then he flew me there, I did an on-air shift, and got hired.

I was probably a little old to be starting my career as a night show guy in Chicago; I think I was 30 at the time. I was feeling a little bit frayed on the outside … I was probably stretching the outer end of the demo for myself there.

SLUDGE, *Air personality*

I remember him coming in; he was on after me. We kind of hit it off right away. He was asking me how I put things in the system, how I set up things. He instantly started asking me how I did some of the things he was hearing, which was a big compliment. I love Fook; I think he's a great talent.

FOOK

(Sludge) was the guy that I first heard when I rolled into town, and my thought was, "I will never be as good as this guy." It was Sludge's old chestnuts of girls and summer dresses, and I was just, like, this guy's fucking great.

SLUDGE

That's amazing he said that. Later on, I remember telling Dave Richards at one point, probably eight months in; I was, like, "Fuck, man, Fook is keeping me on my toes." He was such an amazing show that I felt like, "Fuck, I've gotta step it up."

ALEX QUIGLEY

I was still in college, so I was in awe of everyone and everything. This was the great Sludge, this was the great Ned Spindle, and this was Robert Chase, who was the most laconic, laid back, cool dude you could ever want to be. And Dave Richards, who I think is going to end up being the hero of this book, when it's all said and done. I think he was larger than life; he even sounded larger than life. But I was 22. I was wide eyed; I wanted to do every shift; I wanted to do everything they asked me to do. And I couldn't believe I was getting paid, like, Chicago union rates, while I still lived in an apartment in Champaign. I did one remote. I bet you everyone passed on it. It was something at Navy Pier. The remote rate was $500, then, which was more than a month's rent down there. For two hours, I worked. So I went out there, shook every hand, kissed every baby, said, "Q101, yeah!" I remember the AE was like, "Wow Alex, you're really into this." And I was like, "Damn right, I'm into this!" I went back to Champaign, rented out a bar for all of my friends, and had a goodbye party. It didn't even cost all of $500. It was unlimited, "Whatever you guys want." And I was thinking, "Holy cow. This is what life is going to be like?" It was—the culture was—I don't want to say, work hard, play hard. It was, like, work hard, work hard. Compete with each other, but remember that we're greater than everyone else. So, it wasn't like, slag somebody, like, your fellow on-air talent. It was, "Shit, I gotta get better than Sherman, 'cause he sounded really good." Or, "I need to make sure that I'm the one person they want to call when Robert goes on vacation for a week." It was competitive, but a healthy kind of competition.

SHERMAN, *Air personality*

It was April 15; it was Tax Day 2001. That was the first day that the then-Program Director Dave Richards put me on the radio. It always just sticks out because I was, like, "Oh that's fitting; it's Tax Day. Perfect."

It was totally different from what I was used to, from where I was before. I had started at a station up in Kenosha called 95 WILL Rock. And nothing against that station, but it was not a technically major market radio station. So you'd do smaller market things. When you did a big show, your big show was at the Brat Stop. No offense to the Brat Stop, but the Brat Stop holds, give or take, 400 people, maybe. I came into Q101 and it was a totally different animal. I use the analogy of being brought up to the major leagues in baseball. All of a sudden, all of the baseballs are all white all the time; there are no dingy baseballs. The studios are beautiful. When you go to do a stage intro, you've got 19,000 people in front of you. It was a totally different animal. Granted, it was the same thing. Dave Richards actually said it to me … he went, "Radio is radio," and he was absolutely right.

FOOK
Sherman hates me. I think Sherman hates me. I used to make fun of him for being from Wisconsin, but it wasn't because I hated Wisconsin or because I hated him. I don't know why I did it, I guess.

KATIE SHERMAN, *Senior Project Manager*
I became roommates with the newly promoted overnight guy named Sherman. Lived with him and another part-time DJ. Sherman eventually became my best friend, then my husband, and now we have a child together.

Sometimes I accidentally still call him "Sherman" instead of Brian, because that's how everyone refers to him and that's how I knew him before we even started dating.

ALEX QUIGLEY
(Sherman was the) glue. I mean, he was there for ten years, man. And he stuck through it … He was the kind of person you wanted to have work at your station.

SHERMAN
I started overnights because Alex Quigley left; he went to Bakersfield to some kind of Program Director job.

NIKKI CHUMINATTO, *Air personality*
(Sherman) was one of the first people I met. Super nice, and each time he would move up the ladder there, you had to get excited, because he worked his ass off.

FOOK

Dave Richards was canned six months after I got there, which I found odd. For whatever reason, the station sort of continued its momentum for a year after that and that's kind of when the wheels came off the cart. What happened to alternative music and alternative music radio stations, sort of symptomatically across the country, caught up with Q101. That was really the beginning of the end, I think, a year and a half after I got there.

ABE KANAN, *Imaging/Producer*

Dave Richards, he's the only one. I thought he was great. When they got rid of Dave Richards, I thought it was the beginning of the end.

DAVE RICHARDS

The General Manager that hired me left; a new General Manager came in. We saw the world in different ways. I didn't necessarily agree with his particular image of me and my abilities. I proved myself before I got there; I proved myself after I left.

CHUCK DUCOTY, *General Manager*

Obviously, there had been some issues or problems, or I wouldn't have been there.

J LOVE, *Producer*

I love (Dave Richards). The guy had a real kind of charisma about him, and knowledge.

MARY SHUMINAS, *Music Director*

Dave Richards worked really well with talent. If you were fortunate to work for Dave, he always took care of his people, and it was great.

DAVE RICHARDS

Q101, it seemed like there were too many people to sign off, and too many opinions were taken into consideration. I'm not the kind of person who does it all by himself and says, "It has to be my way," but when your feet are being held to the fire for performance … but your hands are being held because, "Well, we can't let this person go, and we can't make that change, and what we used to do was this, that, or the other thing, and we have to get a lot of other people's opinions in," it's very hard to operate. It's … what's the best way to describe it? It's claustrophobic.

Early in 2001, Q101 relocated in the Merchandise Mart from its 17th floor facilities, which it had long since outgrown, to a state-of-the-art, second floor, showcase studio.

JEN JAMESON, *Air personality*

They had been building the station downstairs for a while. I remember seeing it when it was just the guts; we'd go down and take tours, and it was pretty exciting. We were so psyched to have big huge studios and TVs and multiple production rooms. It was a weekend, and I remember being the one in the room downstairs, the new ginormous big studio, on the phone with (engineer) Vic Drescher upstairs. I pressed the button to swap the station over to the new studio.

CHUCK DUCOTY

That place was fabulous. I had a friend of mine who came and visited and I took him on a tour of the building. We were down there; he was, like, "This must have been what it was like just before the explosion of Pompeii. These are the last days of Pompeii; this is just fabulous. It was a fabulous facility.

PHIL "TWITCH" GROSCH, *Webmaster*

After being down on that second floor, it wasn't as special or magical as it was upstairs. Maybe that's just my own nostalgia of thinking how great it was back then, but if I had to take the two periods of time … the memories of that 17th floor are more special to me.

ROBERT CHASE, *Air personality*

I remember when we went to the new studio, that massive studio. You could probably fit 40 people in that studio if you wanted. And they'd had a crowd of people in there doing something or another. I don't remember exactly what was going on, but as I drove in to work that morning listening (to Mancow's Morning Madhouse), there was a girl who was pouring hot things on her naked body and they were discussing the pleasure or whatever that came with that. And so, they were done with their show and J Love, I think, said, "You can go in." So, I walk in the studio to get my stuff lined up, and get ready to go on the air, and I hear kind of a moaning and I literally walk around the counter and on the floor is this naked woman playing with herself with, I think, hot wax poured on her or something, and I just was speechless. I was, like, "Show's over. Change over here if you want to …." I was just, like, I went out to get somebody, and said, "Maybe she doesn't understand. There's not going to be another segment."

JIM "JESUS" LYNAM, *Producer (Mancow's Morning Madhouse)*

"There won't be another segment!" That was Chase. That was classic Robert Chase. That's what it was like.

2001: 9/11 & 94/7

CHAPTER TWENTY

In 2001, Rock 103.5 had been off the radio dial for three years, leaving Q101 to target Chicago male listeners without much competition.

In a surprise move, ABC Radio, which had been flirting with various iterations of the "Modern Adult Contemporary" format on its FM station, made a formal format change to "alternative" in September of 2001. For the first time in years, Q101 had a direct competitor in WZZN, "The Zone."

At the helm of "The Zone" was former Q101 Program Director, Bill Gamble.

BILL GAMBLE, *Program Director (WZZN the Zone)*
A lot of people went, "Oh, well, how're you gonna take on Q101?" And it was such a non-emotional decision. At the time, ABC had a single FM. There was only one other owner in the market that had a single FM. That happened to be Q101. Q101 was mired in the low twos, floundering. So when we looked at the market, we said "We have to do something," and Q101 was calling itself an alternative radio station. Meanwhile, it was just all rock music and Mancow … TMX was becoming more and more AC. When we looked at the market, we thought there was a legitimate void for a radio station that could play alternative music. And it would probably be a little bit older; it'd probably go back into the '90s, but certainly try to find some new music that was out there, and capture some of those females that Q101 had. We really considered that Q101 abandoned them; Q101 really became Rock 103.5 when Rock 103.5 became Jammin' Oldies. And we thought that was our best chance of winning, because we knew we couldn't win as a classic rock station the minute Bonneville decided to take on The Loop. We knew Bonneville

would out-spend us to death; so, we knew that wasn't an option. So, that's where it went.

And what was crazy about it was, we assumed, and this is one of the things you should never do; you should never assume your competition would act rationally or intelligently, because it's a huge mistake to think so. Sitting where we were sitting, we thought, "Okay, you know, Q101, they've got Mancow. Dave's over there programming it; it's gonna be the rock station." Okay, great. We're gonna come on and … it was pop; it had some rock in it, but it was an alternative radio station. It was not like them at all. And we used the "alternative" word. But they instantly did what we never thought they'd do: "Well, we have to protect alternative!" And I have to tell you, of all the things, of all the attributes we thought they would try to go for, we thought they'd try to protect rock, they'd try to protect Mancow, they'd be the "guys station" with Sludge and all the guys over there. They didn't. They protected alternative. They started fighting tooth and nail for alternative. They started going the other way! And so, you know, we did this for about a year, and it was, like, "All right, what the hell?"

CHUCK DUCOTY, *General Manager*

The week before 9/11, WLS (sic) announced that they were going to go alternative, and then 9/11 happened. It was pretty much the perfect storm. It wasn't necessarily what I thought I was signing up for, but shit happens. It is what it is.

We had parted company with Dave Richards … I was literally the de facto programming guy. When you have a talent like Cow, you really need to pay close attention. My life was getting up at 5:30, monitoring the show, if need be getting in touch with Jim, the Executive Producer, and making sure we didn't go off the rails. The morning of 9/11, I got up and turned the station on at 5:30 and Cow and the crew were setting up the day's bit. The day's bit was going to be that they were going to play a hoax on the audience at 8:00. If you were listening before 8:00, you would know it was a hoax. If you were listening at or after 8:00, and had not been listening before 8:00, you would not know it was a hoax.

DJ LUVCHEEZ, *Technical Producer (Mancow's Morning Madhouse)*

Every now and then, Cow would say something between 5:30 and 6:00, and say, "At 9 o'clock we're going to say that Turd, blah, blah, blah, something," so people who know (earlier), call back in then. That morning, Cow was like, "We're going to do a hoax that's just going to freak everybody out. When you hear it, you're not going to believe it." None of us knew what it was; we just knew that Cow was going to do something.

FREAK, *Air Personality*

The way the studio was situated, the screens were in the corner, and I would just watch TV while the show was going on, just for that reason—breaking news. It happened just after the top of the hour, and we were getting ready to go to our break, when they broke in with the first tower. I interrupted us going to break. I was, like, "Hey, look at this. A plane flew into the Trade Center." Then we watched that for a little bit. And we were like, wow, and we really did have to break; it was like seven minutes after already; we were already running late. So, we were, like, "We'll keep our eyes on this. We'll take a break; we'll be right back." Mancow was just about to hit the button. I was staring at the live coverage, and I saw the plane fly right into the building and all the flames shoot out the other side, from the other side, and I just screamed, "DUDE!" That's what came out of my mouth, the word DUDE. I was, like, "I just saw it. Another plane flew into the tower." Cow hit the button that brought the FOX News feed back live on the air, and you heard the announcer say, "Well, apparently another plane has flown into the Tower, and this doesn't appear to be an accident."

CHUCK DUCOTY

My immediate response was, "What the hell are we doing now?"

MIDGE "DJ LUVCHEEZ" RIPOLI, *Technical Producer (Mancow's Morning Madhouse)*

Everybody thought it was a hoax at first.

CHUCK DUCOTY

I'm driving in, and it is a couple minutes after 8:00, and they are now talking about a plane that has crashed into the World Trade Center. I'm thinking, "Why do you guys think this is a good bit?" This is the time they were supposed to do the hoax. I'm thinking, "What the hell are you guys thinking?" I'm starting to play all this stuff through my head—what are we going to tell the press? What are we going to tell the FCC? How is this going to play? I'm also thinking, "They are doing a really good acting job on this." I remember Cathy Vlahogiannis ("Freak's Niece") was still on the show, and I'm thinking, Cathy sounds really believable. They were all saying they were watching the monitor.

FREAK

This is going to sound weird, but I've had a couple people tell me that, to them, I'm like this generation's "Oh the humanity guy" with the Hindenburg. Because we were at the top of our game, we were syndicated, and a couple million people were listening, and I was the one that yelled, "Oh my god. It just happened. Another plane just flew into the building." Just this 9/11, I was sitting in the bar and this dude walked up to me and was, like, "Dude, you were the one who told me about it." A lot of people were listening, and more than half of them thought we were making it up.

CHUCK DUCOTY

I thought, let me just see what's going on. I punched up (W) GN. 'GN was talking about the plane that hit the WTC. I'm going to reveal my own character defects; I'm not proud of this, and here's a study in human nature: when push comes to shove, it's all about survival. And I will tell you there was ... my first momentary reaction—I'm not proud of it; it didn't last very long, maybe a nanosecond—my first reaction was, thank God, it's not a hoax. And then, I'm like, "Oh, shit, what did you just think?" Oddly enough, there was a moment of, "Okay, it's not a hoax. Then it was, "Okay, now what the hell are we going to do?"

TIM JOHNSON, *Marketing Director*

Suddenly the show exploded in my ears as everyone gasped or shouted. Mancow, obviously shaken, began explaining that a plane just flew into one of the World Trade Center buildings. He described what he was seeing on the TV in great detail. The smoke, the fire, the debris falling.

BRIAN "THE WHIPPING BOY" PARUCH, *Air personality*

I remember standing by one of the offices, whichever direction faced the Sears Tower, continually looking at the Sears Tower, thinking (that) any second a plane could hit that building.

CHUCK DUCOTY

When I went in, we let the staff go home and they did have the choice. Robert Chase, who at the time was doing mid-days, said, "I want to go home." And he went home, and I think J Love filled in for him.

ROBERT CHASE, *Air personality*

I chose not to be (on the air). I arrived at work like everyone who was up and was aware of events taking place that morning—heard about them, saw them, arrived at work about 9:30, I think; that's about an hour—the towers had not yet fallen when I arrived at work. I remember standing there with the whole Mancow crew. I remember Midge, and Jim, and J., when the first tower fell. That was just a moment, obviously, I can't forget. I had a 2 ½ year-old at home at that point, who had a doctor's appointment that morning. I remember the General Manager, what was his name at that point, Chuck? Not Hillier, but the other Chuck. And he approached me because I was getting ready to go on and he said, "If anyone wishes not to stay here, you're free to go home." I opted to go home to be with my wife and my daughter; it just felt like a family thing had to happen. And really, of the staff, I was the only one in that boat. Everybody was free and single or married without kids. I was feeling the weight of wanting to comfort (my) family really, and just absorb this, and radio seemed pretty insignificant to me that day.

CHUCK DUCOTY

Honestly, I gave everybody the option. I really didn't expect anybody to say, "I want to go home." Honestly, that was very disappointing.

TIM JOHNSON

Robert Chase, who usually followed Mancow, stated that he wanted to be with his family and left the building. Mancow was wrapping up his broadcast, telling his listeners that he was leaving the building and implying that it wasn't safe. Who was going to take the controls on the air? What should we be doing? News? Music? Automate? There wasn't a plan for anything like this.

J LOVE, *Producer*

I stayed on and ran the ship the entire day.

BRIAN "THE WHIPPING BOY" PARUCH

J Love and I stayed on the air for like six or seven hours that day after the Mancow show as, like, the guys on the air …

J LOVE

Everyone left. I stayed and ran Q101, at least the board op portion of it. Whip and Freak stayed too, because it happened in the morning and we were just kind of the guys that manned the ship for most of the day.

TIM JOHNSON

The office was thinning out quickly. Since there wasn't a PD, Chuck huddled with those left and came up with a strategy to keep two of the morning show characters on the air to play music and provide updates. The rest of us remaining were tasked with continually finding out as much as we could to feed the two hosts for each break.

I sat at my desk, TV blaring, typing any and all information into a memo that would be passed into the studio for "Prison Bitch" (J Love) and "Brian the Big Gay Mule" (Brian "the Whipping Boy" Paruch) to read. I was in shock. After twenty years of broadcasting, I was stealing from the talking heads to feed two sideshow sidekicks on one of the biggest and most important radio stations in the U.S. This was not what I thought I would be doing during a national crisis.

FREAK

I just stayed on the air playing music; it was me and Brian. We had MSNBC and CNN and FOX, and we were just watching TV. We'd play like three songs, give an update, play three songs. That's how I spent 9/11, even though they came in and they told us they were going to evacuate the building. I was just, like, this is my job; this is what I've gotta do. The security guard was like, "Well, you gotta go." I was, like, "You know me; I'll be back in the studio before …." He went, "Yeah, yeah, yeah. All right. Well, if anybody asks, I told you to get out." So, they left Brian and me in there.

J LOVE

It had been hours since the actual attacks. Chuck DuCoty popped his head back in the studio, and he had three Subway sandwiches. He said, "You guys are doing great, I'm listening; it sounds good. I thought I'd come back and bring you lunch." And I'm thinking, "Subway's still open?"

JEN JAMESON, *Air personality*

Whip and Freak stayed, so they started calling me by, like, 11:00, "Like, can you get here right away? You've gotta come; they've been on the air since 5:00 this morning.

SLUDGE, *Air personality*

I was not on the air on 9/11. I took a vacation to go see my parents in Toledo. It was all going down there, and my first instinct was to drive back for the four-hour drive to Chicago and get on the air—not in any kind of glory hound kind of way, but I always felt like the listeners—and this is probably not a healthy attitude—but they're like my real friends. I felt, like, I wanted to be there with them when this event was going on. Friends in Chicago were calling me and saying, "You're not going to believe; the Hancock building is a target." I was, like, "What?" I was so fearful for people there, too.

FOOK, *Air personality*

That was one of the weirder experiences I've ever had. That was a few months (of me) being in Chicago. I remember waking up that morning. I had rented a place on Lake Shore Drive, and my parents woke me up saying, "Dude, turn on the TV." And through the whole day—I think I got a call after the whole thing happened, and somebody told me that I could either work or not work that day because there were warnings for all of the buildings downtown. I said that I would work. I really wasn't afraid for my physical safety, but as the day wore on, I became more and more afraid about what I was going to say or do. Thank God, I wasn't a talk show host, but the idea of trying to comfort people who were actually listening to you or say something properly to somehow encapsulate the moment—even just play the right songs and not play the wrong songs—was really just a nerve wracking task.

I remember driving downtown and the city being absolutely empty, being extremely eerie, and getting on the air and having pretty connected moments to the people who were calling in. There was a little bit of weirdness at first, where somebody had called in and I put him on the air, and he was talking about how he had some Arab guys who worked at his construction site and he was going to confront them over the next week or so. Without being too didactic, I tried to say to him, "Look, this is one guy who's working at a construction site and it's not really about that thing. I would try and let that anger go and feel the grief without turning it into that sort of thing." I got very supportive calls. There were a lot of people showing love and a lot of sorrow that day. I can't say that I said anything that was particularly touching or memorable or so forth. It was just important to be there. It was really hard to crack a mic at that time. You knew whatever you were going to say wasn't going to match the gravity of the moment.

ALEX QUIGLEY, *Air personality*

I'll tell you what, man. (Long pause) That was a rough night. Because that night was just anger. I came in early with Fook. Because I was Fook's ... I don't think there was a Loveline. I think Loveline got cancelled. I think we cancelled Loveline, because Fook was on pretty long, and I was on and ... so, I was on the first night after 9/11. And this was when people had been ... the shock was over, and they were getting drunk and they were getting angry. And every other song, I had to—I made the decision to put the angry callers on the air, but talk them down. Granted, they weren't live, so I was able to edit a little bit, saying, "Hey, don't –" … "I see a towelhead in a cab, that motherfucker blew up my friend!" "No, listen. That guy didn't do anything. That guy didn't do anything." In retrospect – "Ah, we're down in Bridgeview; we saw some of these guys celebrating in front of a Mosque!" And I'm trying to defuse it, like, "You didn't see that. I haven't heard anything about that." "Yeah, I swear it happened, man." Turns out, they found out later, it was the mosque that was part of the ... they had Al-Qaeda connections that were funneling money.

JEN JAMESON

We had to sit and go through the music logs; take out every song that had any sort of reference to killing, bombs ... Drowning Pool ["Bodies"] was popular at the time.

ROBERT CHASE

The next day, I was back at work. (I) aggressively took phone calls and played a lot of phone calls of people's thoughts and concerns, trying to comfort each other, and (I tried to) play music that maybe addressed the situation. I just tried to work through it, because it was, like, the show must go on. This is what I do; this is odd;

this has never happened before. We're all just kind of going to work through this. That was 9/11.

SLUDGE

I felt so weird when I got back, days later. I remember playing "God Bless America" on the radio, that old classic version. Al Roker, Jr. walked in. They were all listening in the other room, and he came in and went, "We were all listening in the other room. We're, like, "What the fuck is Sludge doing? Isn't he a little late to the trigger on this one?" Then we all remembered, "Aw shit, he's just getting it for the first time; so, he's gotta re-set up everything."

EVERY DAY IS EXACTLY THE SAME: MORE CHANGES

CHAPTER TWENTY-ONE

TIM RICHARDS, *Program Director*
I started in October/November of '01, and they let me go in July of 2003. The environment (at Q101) was kind of a mix. There was an incredible level of professionalism there, at one level. And then there was the craziness that was Mancow and his crew that definitely kept things interesting on a daily basis.

RYAN MANNO, *Air personality*
My internship wrapped in August (2000), but then I moved right into interning for Sludge. Then I got hired as a P.A. (Promotions Assistant) following that, working in the promotions department, until … Tim Richards hired me on the air. I walked in; I was doing a college radio show at the time at North Central College in Naperville. I was hosting a show with my brother (Kevin) and Gordon (Mays). It was a punk rock radio show called "All Ages Radio." I also did Monday, Wednesday, and Friday afternoons or something.

I took in an aircheck to Tim Richards and said, "This is probably really inappropriate, but would you mind listening to this?" I don't mean to sound cliché, but it was one of the greatest days of my life. He put it in, listened to 30 seconds, and said, "Can you start on Saturday?" I was a junior in college.

He didn't even make it through the whole three-minute aircheck. He listened for 30 seconds, stopped it, and said, "Welcome aboard." I was basically looking for feedback as a PD, "What can you tell me? What can I improve?" That changed my life.

JIM "JESUS" LYNAM, *Producer (Mancow's Morning Madhouse)*
Ryan was young, and really into the local music.

RYAN MANNO
I was disappointed (my first time on the air), and I mean truly disappointed, that I could look at the logs. I didn't understand the concept that … college radio had some programming, but for the most part, we were freeform. That part was disappointing, walking in and seeing that I could tell you what was playing at 3 o'clock next Thursday.

When I got in there and did my first break, for whatever reason, whoever handed it off to me—we were going through some buttons one last time. I had a good handle on everything. Well, I left it in "air." So, there's a program and an air feed. I left it in the air feed, which I think was an eight-second delay multiplied by two; so it was, like, a 16-second delay.

Green Day was ending; "Brain Stew" was the first song. I had cracked the mic for the first time. I felt like my face was going to turn into wood. My parents were listening … it was an overnight; it was so lame—my parents set their alarm—it was a big deal. I think my first words—it was "Brain Stew," and I said something about "a stew of brains." I could hear no words coming over my headphones; so, that was all I said. I turned even more pale than I was. I turned the mic off, and so, I think my first words ever on Q101 were "a stew of brains." It was a disaster. It was horrible.

SLUDGE, *Air personality*
I remember Manno coming in to the station his first week; he was a real musicologist—really knew everything about bands and new music, so he tested me right away with an attitude almost. Instead of saying, "Who the fuck are you? Shut up," I wanted to talk to him about it.

RYAN MANNO
(Sludge) taught me so, so, much about how to carry yourself as a radio host. And he was genuinely so giving and accommodating. One of the first things I did with him was a live broadcast at Twisted—8, maybe? He had me backstage

at Twisted with him, and he put me on the air; he was just very encouraging of young talent. A lot of guys, they get there and they don't even want to think of relinquishing their spot, in a sense, and he just saw any element that he could, and he thought, "How can I make this interesting?" I really do value that time with him, probably as much as I value anything at Q101.

ROBERT CHASE, *Air personality*

(Tim Richards) had come on board as a new face, making (him) my fourth program director. We never talked; he never came to me. I would say hello and good morning and fish around for some sort of back and forth to get together and talk about what (his) expectations were. You're just curious, a new guy, "What do you think of me? That never happened, so I began to see the writing on the wall that either it's A) I'm fine or B) I'm not fine, and it turned out to be B, that he had plans to replace me with a person who had been his music director, I guess, whatever her name was. There was probably some salary attached to that; I was approaching ten years. With ten years I think came a bump and some significant something or another.

TIM RICHARDS

(Robert Chase) was a great jock, but it comes down to vision, and the vision that I had for the station was a little bit younger, a little bit hipper. At that point, to me, I didn't think that was the sound that I was looking for from the station, long term.

ABE KANAN, *Imaging/Producer*

Tim Richards: what does he do? He fires Robert Chase … he brings in Nikki.

JEN JAMESON, *Air personality*

Tim Richards came in, and I was on vacation with my family. (I) came home, (and) had a meeting set up with him. I was so excited to meet him. I get there and Mary (Shuminas) gets pulled in to the office, and he fired me … me and Robert Chase. He felt the need to make some changes as the new guy. I met up with him (at) South by Southwest four or five years ago, and he apologized. He said, "I didn't know what I was doing back then. I felt like I was forced to shake things up a little bit. I'm sorry. I didn't mean to let you go like I did." The day I got fired, I walked out of the office; I grabbed my stuff, I walked to Baskin Robbins, got an ice cream cone, and left.

SLUDGE

Chase was awesome. Chase was like Jello; he was so relaxed all the time. We had a lot of good music conversations because I played guitar and he did too. He'd try to introduce me to bands; sometimes those bands were pretty out there. I'd nod my head and smile and pretend to listen and like it. Robert was probably the nicest guy in the building. Just a super laid back, chill, hippy dude. I always thought he'd end up on XRT, and be a long-time personality on XRT. Instead, he's in Montana, loving Montana, right now.

ROBERT CHASE

Honestly, I think I was probably ready for something new, too. Having been there nine years (and) having been through all the alternative music that we brought, (which) was kind of waning, we were fishing. (We were trying) to find the right mix of music, I think—to be as relevant as maybe we had been in the mid 90s, and just trying to figure out a way to crack the ratings. I think we were struggling there for a stretch, and maybe I was just a little past that demo, admittedly. And he said, "You're done. Thanks for your service here;" it had been nine years, just a couple of weeks shy of a full nine years there. I appreciated it, and walked away, just looking back going, "Wow, nine years, man, that's okay for a kid that was just a small-town guy hoping to clock a few months and say I did it."

TIM RICHARDS

(Nikki and I) had worked together in Detroit, and she was definitely more of a rock-type personality. And we had a great working relationship. I wanted to have somebody with an additional set of ears on the product. She came in; she was AMD (Assistant Music Director), and she did mid-days.

NIKKI CHUMINATTO, *Air personality*

I had come in, I would say in October or November, to go see Weezer at the United Center, and I loved Q101. I remember driving in, and it was a big rock station; it wasn't the alternative station that I had remembered from my prior visit. SO, that being said, I went to Weezer. Whoever did the stage intro got booed—I think it was Sherman, but I can't confirm that.

SHERMAN, *Air personality*

There was a stage intro, I remember (music journalist) Jim DeRogatis mentioned me in the paper one time because I got the loudest boo he had ever heard in his life doing a stage intro; (it was) for Weezer in 2001. In 2001 when I came into Q101, it was cool for people to hate Q101—especially Weezer fans, because they hated anything to do with corporate, so they looked at Q101 as a corporation—which turned out was true, anyway. So I was a newbie—I had done stage intros at the Brat Stop (in Kenosha, WI) before that. You didn't really have to have a plan; you just went out and (said), "Hey, I'm from the radio station," and expect them to cheer. You're in the United Center, you've got 22,000 Weezer fans, and they just booed off Cold. They just booed them off; so, I'm going to come up there and go, "Hey, I'm Sherman from Q101." What do you think's going to happen? They just booed me so loud I was like a heel in the WWE. I just held up my hands and (went), "Come on," and let them just boo me even louder. They booed me so loud it made the paper.

NIKKI CHUMINATTO

I came back to Detroit, where I was working under Tim Richards at (W) KQI, and I was, like, "Oh my God. Q101, this station that I loved, is totally different. I don't even know what they're doing. It's like all metal; it's got nothing to do with the Weezer and Liz Phair, and all that stuff that I grew up with and loved." That being said, he took those comments, hired me, and then brought me in, in February, thinking I would try to change it to the Q101 that I described.

RYAN MANNO

I loved Nikki. Nikki was that kind of chick that you felt like was out at the show the night before. You could hear almost as if she (did) not just drink the bottle of whiskey, but ate the glass. You felt like she was living the lifestyle, and she was.

NIKKI CHUMINATTO

I lived it up like crazy. I didn't have any kids or anything at the time. I was probably out four to five nights a week. I got to see some of the greatest shows I've ever seen in my life. I developed great friends. I still talk to Mary; I talk to Nicole (Claps-Gamboa, future Music Director), and I talk to Tim.

DAN "BASS" LEVY, *Producer*

It was so much of a family organization; everybody knew each other. When somebody new came in, you met all the sales people; they were all cool. It just seemed like everybody got it, as far as radio was concerned. Everybody knew that it was going to be fun; everybody knew there was business involved; people were making money. There was just this creative atmosphere. The lineup of people that were on the air then was Mancow, Nikki, and then Sludge, Fook, and Sherman. It was just like this energy. Every show was so creative, one after another.

TIM RICHARDS

The worst thing I could do would be to say that it was the environment's fault for the mistakes that I made as a Program Director. But there was a certain level of toxicity to the environment, for me, being relatively young in the business and definitely not ready to program a station like that at the time; it was a huge learning process for me. The station's staff, the whole alternative mindset, was a lot different from the happy-go-lucky top 40 world that I came from, coming back to Chicago.

NIKKI CHUMINATTO

It was exactly what you'd expect when you work after people named Turd and Freak.

RYAN MANNO

People always say you never want to meet your heroes because they'll disappoint; this is such an exception to that rule. At that time, it was as crazy as I'd wanted it to be. It was as unique and alive as I'd hoped that it would be. Walking in, you didn't know what you were going to see … midgets stapling money to each other's heads.

NIKKI CHUMINATTO

You really don't even think that it's real. You really don't. You hear that "there's a stripper in the studio, and we're all drinking Dos Equis," and you're, like, whatever. It's all just radio theater of the mind. And then I got there, and I really did see boobs and beer in my first week. It was hilarious. I'm not easily offended. I think if you are easily offended, A) you shouldn't be at Q101, and B) you shouldn't really be in radio. I kinda thought it was funny. Welcome to my job. I like beer, and I've got boobs. I'll fit in fine.

CHUCK DUCOTY, General Manager

Somebody in your interviews may tell you otherwise, but I think we got the overt drinking in the halls under control. Did we have strippers in the back? Yes. Did all the things that come with that kind of a morning show still exist? Yes. Did we try to keep it confined to the studios and the area in the back, and hopefully insulate the rest of the staff from it? Yes. Were there occasionally some problems with somebody coming in to the station in the afternoon, who had had a little too much to drink during the day, (who came) back, and was disruptive? Yeah, that happened a few times. And you know what? A couple of times we suspended people because of it. I think you still have to maintain a modicum of control, but again, it's the definition of what that control looks like. To me, part of that morning show was you're going to have strippers. They're going to be in the building. Let's try and confine them to an area where they're not disruptive or a problem for other people. You're going to have "Drunk Girl Friday;" somebody is going to be drunk on the air; it is going to happen. My job was to protect the license, and when you have a morning show like Cow's, that becomes probably a little more problematic.

NIKKI CHUMINATTO

Six months in (to my time there), the music was primarily alternative. Toward the end, it got really pop-punky, like Good Charlotte-sounding.

ALEX QUIGLEY, *Air personality*

Right before I left in '02, I forgot her name—oh God, the record rep for The White Stripes ... She played "Fell in Love with a Girl." And then, like the very week afterwards, it was "Last Night" (the Strokes), and then it was Thursday, "Understanding in a Car Crash." And it was—I was starting to see, you know, because I was doing overnights, I played all the "C" and "D" (music rotation category songs); like, I played all the new songs. I was, like, "All right, this rap-rock bullshit is over! And these songs still rock."

NIKKI CHUMINATTO

The first Jamboree—nothing ever really gets you ready for that. Twitch and I were the video hosts. I had a Boone's Farm shirt on and a cowboy hat—things I would never do in normal life. Just walking around and running into Jack Black and then the Strokes, it was awesome.

DAVE BALL, *Promotions Assistant*

For me, Jamboree was always the most fun. When you're a big fan of the music that you're a part of, you get to meet all these really great bands. You get to meet bands like Silverchair. I'm like, "I grew up with Silverchair,—I get to meet those guys now?"

CHUCK DUCOTY

The last Jamboree we did outdoors (2002 – Dashboard Confessional, Earshot, Hoobastank, Kid Rock, Local H, Our Lady Peace, Quarashi, The Strokes, Tenacious D, Thursday, Trik Turner, Unwritten Law, X-Ecutioners, Zwan) was probably the least successful one we had, but the concert business was changing pretty dramatically at that point, too.

WE ARE ALL MADE OF STARS: MUSIC IN THE 00s

CHAPTER TWENTY-TWO

By the end of 2002, Q101 was working to reclaim its "alternative credibility" by edging itself away from the knuckle-dragging artists associated with the so-called "rap rock" genre that exploded around the turn of the century (Limp Bizkit, anyone?). While the station worked to resolve its identity crisis, it took a credibility hit from a national source: Entertainment Weekly *magazine.*

From the January 3, 2003 issue "Loser of the Week: Chicago's Q101. Responding to news that Stereolab's Mary Hansen had died in a traffic accident, Fook, a DJ on the pop station, played a sound effect of a crash and said the band 'sucked.' It's a shame the FCC can't regulate class."

TIM RICHARDS, *Program Director*

It was an insensitive comment about the passing of the lead singer.

FOOK, *Air personality*

What was the quote in *Entertainment Weekly*? It was Loser of the Week, then it was Chicago's Q101, then it went on to talk about me and it had a picture of a cartoon me.

I was in the wrong; there's no two ways about it. What's funny about my experience at Q101, or in Chicago, is that that was my first lesson that I really had to watch what I said. Before that, I was unaware, because I was in a variety of other markets where I said what I said, some of it for shock value, some of it was legitimate humor, (and) some of it was legitimate commentary and had thought behind it. That one wasn't. The funny thing about Stereolab (was) that I had zero idea that if Stereolab did have kind of a home in the U.S., it was Chicago. Considering Q101's format at that time, I still think, probably, that 99 1/2 percent of the listening population had no idea who Stereolab was. From how I understand it, a writer from an independent Philadelphia newspaper, or maybe he was from the big paper there, but he was a music critic—he was passing through Chicago at the time, heard it, got upset, and broke that I'd said this, and it traveled along message boards, and eventually, blew up into what it was.

TIM RICHARDS

Fook's a talented guy, but not necessarily the most compassionate guy on the planet. It was a moment to be cool, which was his job description.

FOOK

As far as the actual comment itself, (it was) completely unwarranted. I mean, I had heard Stereolab for the first time that evening, took a clip of them off the internet, said that it sucked—I didn't like the clip that I heard, but that's no excuse—and it's a moment that I'm 100% not proud of. But it was a real lesson to me of the power of the radio and that people are occasionally listening.

It happened in middle December, I think. I remember being in Chicago over New Year's, just kind of huddled in my apartment, looking at that *Entertainment Weekly* article, and being, like, "What the fuck is wrong with you, man?"

NIKKI CHUMINATTO, *Air personality*

Fook and I would meet for beers every once in a while. He's a good guy. How would I even put this? Just totally different. What was it, the Stereolab thing that got him in trouble? Just a totally different sense of humor, but not a bad person.

FOOK

I feel terrible about it. The idea is, look, as a human being, why would you ever disparage someone's death that you didn't even know. It was something so stupid, that disc jockeys do all over the world all the time, and only a few of them get called out on it. At this point in my life, I'm happy that I was called out on it; it made me a better person. It was just a shitty thing to do.

AJ COX RUDLOPH, *Sales Promotions Manager*

I love Fook. Fook was a little over the top sometimes. He had a habit of yelling at interns, like … wow. You definitely could walk down the studio hallway sometimes and hear the studio door fly open and Fook yelling something; and then, the intern would walk out and the door would slam behind, and the intern, more than once, would go, "Fucking Fook."

KEVIN MANNO, *Air personality*

There were times when he did make me feel like shit, but I sort of expected it; that's like the stereotypical intern thing, but Fook and I always got along. I just kind of stayed out of his way, and he would have tiny little tasks for me, like, I would go get beer and hot dogs at CVS while he was on the air. Then I would do the hot dogs in the microwave, and we would drink beer and eat hot dogs.

MICHELLE RUTKOWSKI, *Air personality*

I have so many great memories of working with Fook. Some of the dumbest and 'funnest' things I ever did at the station were under his tutelage. I dressed up like Jesus and handed out Valentines on the "L." I put on skis and waited for someone to come find me in front of a sausage shop. I got carried around on the shoulders of a bunch of drunk dudes at the Southside Irish parade. For two years, he hosted "Fook's Footie Pajama Slumberama" where we had bands (Vendetta Red, Riddlin' Kids) play in the studio, and then listeners won the chance to come and see the bands, but then to have a slumber party at the station. Giving people that kind of access to a radio station was pretty special—and a little stupid, if you think of it, from a legal perspective.

FOOK

Hey, I don't think (I was difficult to work with). I had a lot of interns; it was kind of the job of the night guy to have interns. I was certainly harsher on my interns than other people, and I'm pretty sure this is where I got the reputation. I guess I was probably 'weirded out' when some things at events that I was hosting didn't go so well, and I probably went to task on some promotional staff. I doubt that's where it came from.

RAVEY, *Air personality*

I thought Fook was a lot of fun; he always really supported us. I think we made real fast friends with him. He was always great about telling us if he heard something on the show that day that he really liked; he would be the first person to step up and say it.

NATALIE DiPIETRO, *Director of Fun and Games*

Fook is extremely talented and I always felt badly for him because there was so much attention put on Cow, and then Woody, Tony, and Ravey, that sadly, I think Fook's talent got overlooked a lot.

SHERMAN, *Air personality*

There are a lot of words that come to mind with Fook. Fook and I had our differences. All in all, he's a good jock. I would say … what's the word I'm looking for here? Arrogant.

RYAN MANNO, *Air personality*

I love Fook, but man, he was such a bitch. He really was so difficult …

FOOK

I think if you talked to a lot of my interns today, you'd find that I'm friends with a lot more of my interns today and that I do take responsibility for my attitude at that time.

SLUDGE, *Air personality*

It had to be probably ten years ago. Bands used to come in and play for the staff before their record came out and before they were anybody, just to help promote the band.

Maroon 5 came in to the conference room, and played acoustic, a couple of songs. This was before their record came out; before anybody knew who they were. I actually liked the music a lot, and started talking to the band afterward, especially Levine.

We just kind of hit it off, were very friendly to each other. We started talking for a half hour. Finally, he said, "Look, we're not playing anywhere tonight, but I want to see the town a little bit. You wanna get some beers, hang out, and you could show me Chicago?" I said, "Yeah, that's great. Let's hang out," and so he gave me his cell phone number. I said, "Yeah, I'll give you a call later. We'll party later." He said, "Cool, looking forward to it, because I don't know anybody here."

I did my afternoon show, and went back to my apartment, which was on Dearborn at the time. I just was like, "God, I'm just tired. I'll just take a nap; I don't feel like going out tonight and running around." I fell asleep and never called him.

Six months later, I was watching the *MTV Movie Awards* with some Q101-ers. We were watching it, and all of a sudden, that "Harder to Breathe" song comes on. I go, "God, this song is awesome; who is this again?" I think it was Jim ("Jesus" Lynam) or someone who said, "You know; it's Maroon 5. Remember? They came in six months ago, the singer wanted to be your best friend, and you blew him off? Now he's the biggest thing ever, and he's banging Brazilian supermodels and you could have had his leftovers."

NIKKI CHUMINATTO

Chris Martin and Jonny (from Coldplay) came in and hosted "Nikki's Nineties Nooner" with me, and that was the day that Nirvana released "You Know You're Right;" it was an unreleased song by them. They sat on the show with me, and hosted it. I was, like, "Hey guys, you wanna hear this song?" They were super excited. We all put our headphones on, we listened to that song, and we talked about it after. Years later, like seven years later, I saw Chris Martin, and he was, like, "Oh, my God, you're the one who played me that Nirvana song." He totally remembered that.

SHERMAN, *Air personality*

I almost punched out Dryden from Alien Ant Farm because he hit me in the kneecap with a microphone when I intro'd him one time. Had he not had a seven-foot tall, 400-pound bouncer, I would have decked him on stage. He didn't want me intro-ing him anymore, so he pulled the mic wire. I was holding the microphone, and when he pulled the wire, the mic flew down and hit me right in the kneecap; they started playing immediately. I went back to the radio station. I tore them a new asshole for probably ten minutes on the air, and from that point, said, "I'm never playing an Alien Ant Farm song again."

NATALIE DiPIETRO, *Director of Fun and Games*

A much better memorable experience was an intimate performance we did at the very, very, tippy top of The Metro with Beastie Boys. I never even knew there was a very small performance area up there and I had lived in Chicago about 13 years. I remember the stairs being very narrow to get up to it and nearly killing myself running up and down them that day. We all got our work out that day, that's for sure and to see an artist that big in that atmosphere was incredible.

SLUDGE

Radiohead came into the studio and took over my show for an hour. They never did that. I don't think they embraced the whole radio station interview. We let them come in and play whatever they wanted for an hour. I got some candles lit in the studio, and I heard they loved sushi, so I bought some awesome sushi and brought it in. They were really nice; it was welcoming for them when they came in. And they were so rigid; they were pretty serious guys. They didn't like to play around; you couldn't really ball bust and talk about summer dresses with Radiohead. So we started talking for a second, and they were very stiff and they wanted to get going. We were playing the songs they chose on their iPods they brought in. Then, I think I said something along the lines of "Remember when you guys were on the *MTV Beachhouse* playing 'Creep,' and there were all these fraternity meatheads with their hats backwards?" I think they hate that song now, but then, they all kind of chuckled, and it loosened up a little bit. Then, when we started playing a song, a couple of the guys came over and started looking at our equipment asking, "What does this button do?" And I started showing them the equipment, and then they really started laughing and loosening up, and within 10 minutes, we had a blast—off the air, mainly. They really didn't want to talk on the air. Playing their songs, and the stuff they chose, that was their joy, playing this bizarre music. The funny thing was I kept all their water bottles they drank out of. The next week, I gave them away on the air. All 10 phone lines were lit—we gave away a Hummer on the air, a 50,000-dollar car—and the phone lines didn't light up for that as much as (for) the water bottles from Radiohead.

ALEX QUIGLEY, *Air personality*

The celebrity interviews were great. Kirk Hammett (Metallica) was awesome. That was one of the last ones I did. The Corey Taylor (Slipknot) takeover (where Taylor took over hosting Quigley's show) was great, too, because that was something out of the ordinary. It happened in real time, in a way that TV doesn't always—isn't able to do, and the internet isn't necessarily able to do either.

SLUDGE

Courtney Love called up for an interview and she wouldn't hang up at the end of the interview. She stayed on the phone for 2 ½ hours off the air. I had her on the air for 20 minutes, and then she goes, "Well, we'll keep talking." She was talking about all kinds of stuff, about Kurt; I had listeners asking her questions.

I remember I brought in Al Roker, Jr. and Ryan Manno. We all just started talking to her about all kinds of stuff—Nirvana, Kurt, so much stuff. It was amazing. She would not hang up.

RYAN MANNO

On the Bus started in '03. Tim Richards hired me to do weekends and overnights, fill ins, and all that. Part of it was to work with Billy Corgan on air and to be sort of his sidekick/producer of a show called "Radio Live Transmission," which never happened. Billy and I did three shows; I have those CDs somewhere, and I'd love to find them. If you heard them, you would understand. They were horrible. They were awful. So we did three, like, dry runs in a production studio and he didn't like how it felt; he didn't like how it sounded, and he had every right to. It was really bad. It was awkward and awful.

I wish I could remember how On the Bus was pitched. It was basically, like, it was very much that theater of the mind, where for this one hour, from all that imaging that had all of the hydraulic bus brake sound effects to the fact that, quite literally, we were—I would say, 95% of the time—on these bands' buses. I had such integrity for it. I would make a few exceptions, you know: "We can't do it on the bus, but can we come in the studio?" In the studio, it was different. There was something so pure about being on the bus. It was me and a recorder and these bands, and we were where they felt most comfortable; we were in their home. I always loved how things came off. It was interview style, very heavy on talk. The bands would … to tie in music, they would play probably four to five songs they chose, whether it was the Monkees and Jerry Vale, or Red Hot Chili Peppers and the Ramones. The bands would select music that meant something to them. That was interspersed with the most in-depth interviewing that was happening on Q101, or maybe ever. Like, long-form interviews—you know, five, six-seven minutes at a time rather than 30 seconds in between songs.

EVIL EMPIRE: NIPPLEGATE

CHAPTER TWENTY-THREE

Program Director Tim Richards was shown the door in 2003. Mike Stern, who had previously established himself in markets like Las Vegas, Dayton, Milwaukee, and Denver, was tapped to succeed him.

JACENT JACKSON, *Assistant Program Director/Music Director*

I actually interviewed for the PD gig with Chuck DuCoty for Q before they hired Mike. That, in and of itself, I thought, was completely fascinating, because I didn't think that was something I would even get an interview about at all. I just never expected that.

I'm from Champaign, originally. I just kind of heard that they were looking for somebody, and I thought, on a lark, that I would send a resume and all that kind of stuff. I had a little bit of positive press behind me at the time that happened, because the station I was working at—it was in a small market, but it was like a 7-8 share 12+ radio station, kicking ass in ratings. It was enough to get me in the door and talk to Chuck about the job—not enough for me to get it—but enough for me to talk about it. That made me wonder how far Q101 had fallen.

I was really pleased that I got the interview to be the PD for the radio station, but at the same time, I was, like, "Fuck, they're willing to talk to a guy in, like, market #110, about programming this radio station, which means they're completely out of ideas. If that's the case, there's probably a problem here."

MIKE STERN, *Program Director*

Here's what had happened when I got to the radio station. I don't know when this happened and I don't know who implemented it. I don't know if this was a

Tim thing, because he was my predecessor, or if this was a post-Tim thing. I don't know. I'm not avoiding it to be politically correct; I don't know the answer. Here's what I do know: they had done a research study and they had seen that the Zone, which was at that point, fairly new, had not gone full hard rock yet. It was fairly new. Bill had it leaning pretty much towards new music, and they have gotten this research study back that said the Zone was encroaching on Q101's ownership of being the station for new music. The recommendation had been to play more new music—not play the new hit songs more frequently, but play more new titles—so, the playlist was like eight miles long with pretty much every new record that was out there. I think they were literally playing two gold songs an hour, so there were two familiar touch-points an hour. I was up in Milwaukee, and so, when I was interviewing for the gig—this was before Q101 had an online stream—my wife and I drove down for a weekend in Chicago, and essentially listened.

We sat on the lakefront and played cards and listened; we drove around and we listened. Granted, I was programming a rock station in Milwaukee, so I was a little out of the loop for alternative, but I didn't know most of the songs on the radio station. So, that's where it was when I took over.

I remember talking to Rick Cummings and him saying, "What do you think of how it sounds?" And I said, "Do you want me to tell you the truth?"

JACENT JACKSON

I had been a fan of the station. I was a college kid when the station (peaked), when grunge was really burning bright. I was in college during those years; we'd come up and listen to the radio station, and always really admired it. By the time I actually interviewed to program it in 2003, I would listen to it and it sounded like a total disaster. It didn't have any cache. In terms of (me being) a radio programmer, when (record promoters) would talk about their records and say, "Well, we got it on Q101 in Chicago," I would imagine in 1998 that meant something. But in 2003, it didn't mean anything at all, because nobody cared. It was a station that was in a big market, but it wasn't seen as a station that was particularly influential at all, in terms of its cultural impact. I don't think anybody really took it seriously anymore.

JOEY SWANSON, *Air personality*

I … felt that (Stern) made a lot of his decisions to make himself seem decisive. Whether he was changing jocks, adding or canceling shows, changing music direction, it all seemed like bullshit busy work ... like, he was showing the Emmis brass he was doing something, anything, and that he was on top of it.

NED SPINDLE, *Imaging Director*

Dave (Richards) was so hands-off that I could do anything I wanted, and he'd like 99% of it, whereas Mike Stern was hands on, but not in a bad way. Mike got it, and when he made a suggestion, I generally agreed with it.

MARY SHUMINAS, *Music Director*

Mike Stern (fired me). I was done in January of 2004.

AL ROKER, JR., *Air personality*

By the time Mike Stern was there, the whole atmosphere, everything, had changed.

Although I don't have a problem with Mike Stern personally, I really don't, but I don't really have any fond memories. I don't want to say I don't have *fond* memories of Mike Stern 'cause that sounds like I don't like the guy, but, it was just not a memorable tenure.

NIKKI CHUMINATTO, *Air personality*

They moved (Sludge) to mid-days when they let me go. I remember walking down the hallway that Friday, and him just looking at me, looking down. He wasn't like normal, happy, Sludge. I kept thinking, "God, I hope he's okay." And then I got let go, and he called me right afterward.

I went in for an aircheck (with Mike Stern), and left without a job. But that being said, I've been at the Mix now for almost eight years, in February. Eight years in one radio station? That's unheard of. And I'm incredibly happy, and I get along with my bosses, and I have a great time, so I have to write a thank you note, I think.

One-half of one second of a Super Bowl halftime show on February 1, 2004 forced a dramatic change in the way radio operated.

Pop superstars Justin Timberlake and Janet Jackson performed a three-song set during the halftime of Super Bowl XXXVIII (Carolina Panthers vs. New England Patriots). During the third and final song, Timberlake's "Rock Your Body," Timberlake tore off part of Jackson's clothes, partially revealing Jackson's right breast for one-half of a second.

CBS, the network that aired the Super Bowl, was fined over half a million dollars by the FCC (Federal Communications Commission) for the indiscretion. Abject fear of ending up in a similar situation put the entire entertainment industry on notice, once the public became hyperaware of what was being broadcast on the airwaves.

MIKE STERN

That Super Bowl, I was still in the apartment that the company rented me for my move. I threw a little Super Bowl party there.

WOODY, *Air personality*

We actually watched that Super Bowl at Mike Stern's corporate housing apartment. We started the next day. That was our first day on the air.

MIKE STERN

You know, we didn't know; Nipplegate happened, and we didn't really realize what was going to come of that. Nipplegate was such a bizarre thing; it was a television thing—you know. It happened on TV, not radio, and the FCC turned around and started coming after radio for indecency issues. And it really built up over the next... it probably took 90 or 120 days for the FCC to get really ramped up into frenzy. Meanwhile, not only did I have one of the most controversial FCC-censored shows in the history of radio (Mancow), but I'd also added on Woody, Tony, and Ravey in afternoon drive, who were equally, if not sometimes more, blue. So it was a very interesting moment in time. We really had to walk a tightrope ... We almost cut ties with Cow. We almost cut ties with Woody, Tony, and Ravey. It was a real minefield. And you know Woody had outstanding issues from St. Louis, when he had been on the Emmis, KPNT, the Point, there. So, yeah, we were really walking in a minefield.

JIM "JESUS" LYNAM, *Producer (Mancow's Morning Madhouse)*

(Nipplegate) completely changed everything. You couldn't say anything at that point. After that happened, there was really no porn stars in anymore. We'd have the occasional Mary Carey, who was running for California government, and that was pretty much it. The rule became if you were going to have a girl, she was going to have to have something to do besides porn. That was not the rule before then.

ALEX QUIGLEY, *Air personality*

I was in California by that point, though. Boy, that really—look back at that now. It's a nipple, man. I saw—I think it was Drew Magary on *Deadspin* wrote something about how Justin Timberlake got off scot-free. I mean nothing. We didn't even call it, like, Justin—he was the guy who ripped off her clothes! He was the guy who actually committed the crime! Who cares what kind of nipple shield she had on underneath her bra? He was the guy who did it! And he spoke the words, "I'm gonna have you naked by the end of this song." Rip. We love Justin Timberlake! That's an aside.

CHUCK HILLIER, *General Manager*

If Janet Jackson is getting CBS in trouble on television, what in the hell do you think is going to occur at Q101? There was an enormous sensitivity, an overreaction I must say, to shows that were confrontational, or in your face, or had questionable but funny material.

MARV NYREN, *General Manager*

The wardrobe malfunction: it wasn't just us; it changed radio to this day—literally, from that moment on. I think CBS had the Super Bowl that year. The complaints came in right away; the fines were allegedly mounting to millions and millions and millions of dollars. Every CBS station that aired the super bowl was also responsible because that's the way the law was written, at that point time. We had conversations—corporate, management, and myself, about, "Okay guys, what are we going to do?" They were very concerned. I was less concerned because I thought the show that he was doing would pretty much continue as was. It was really more about the FCC and people in the Chicago market, specifically, that were going to say, "I'm going to take advantage of this situation, because now that the Janet Jackson boob popped out, we can now get this guy off the air. We can make sure that our voice will be heard by the FCC." So, I think that was really what started the more in-depth conversations about, "we're going to have to do something" at some point.

WOODY

Everybody's asshole just squeaked shut. It's so funny how far the pendulum swung the other way. I remember being in a meeting with Mike Stern, and he's literally scolding me because on the air—I forget who the celebrity was at the time—I said, "Yeah, dude, I'd totally bang her." He goes, "Really? Bang? We have a target on our back." It's like, really, "Bang?" It was really weird how crazy things went in the opposite direction.

SLUDGE

It was overdone; it was way overdone for what happened. It was just somebody's nipple showing for a second, that probably no one saw until they rewound it on Tivo.

WOODY

If you wanted to talk about how you wanted to murder somebody gruesomely, everybody was cool with that; that wasn't a problem. God forbid you say "titties," or "I'm going to take a piss."

MARV NYREN

Everyone was talked to either directly through me, or through Mike Stern, about "hey guys, we have to be super careful right now." Especially at the early stages, because people were listening to try to catch us doing something wrong. And we were one of those formats—Lite FM probably didn't have to do anything, the country stations probably didn't have to do anything. But if you were a rock and an alternative station, you had to make people aware of what could happen.

SLUDGE

It was more than a memo or an email; it was a meeting with the Emmis lawyers on the phone with every personality in the company in the conference room, talking about the new rules because Janet Jackson's nipple showed up, and "we're in a different world." It was more about "you have to be more careful about your content."

One stupid example was, when we recorded a phone call and played it back. If you bleeped out the word fuck, in the old days, you bleeped out the u and the c and kept the f and the k. Now you had to bleep the whole word out. There were some personalities around the country going, "Well, I have no show anymore; this is great. What am I supposed to do?" I was, like, if that's all you've got, you don't have a show anyway.

MARV NYREN

The edict from corporate to every Emmis market was, basically, a zero tolerance policy, where you couldn't say certain things. The FCC rules and regulations about what you could and couldn't say about excretory issues and … insertion wording, …they were fairly clear, but unfortunately, there was still enough ambiguity, in that there was still enough where you could go down a road, maybe not as far as you would have before, but you could still take that path. I understood in the art side of things, you wanted to push the envelope to make people go, "Oh, man, did you hear that? That was cool." That was also what our radio station was about—kind of cutting edge, being cooler than the station down the street, or down the hall even.

DJ LUVCHEEZ *(Technical Producer, Mancow's Morning Madhouse)*

We were just bleeping everything. I think they totally went overboard. It really took the fun out of radio.

MIKE STERN

I remember every year Emmis had corporate meetings; this was a big deal and every year they had the "Emmys." The Emmys was like a big deal in Emmis. It was (Emmis CEO Jeff Smulyan's) night to kind of celebrate the best of the company. I think for him personally, it was a night of real pride … the night when Jeff really pretty much just wanted to put on his best suit, brag about the best of the best in his radio company, and enjoy himself. We had to sit down in an empty banquet room in a hotel and talk out our strategy for dealing with the FCC and the likelihood of what would happen if we fired those air talents, if we didn't fire those air talents, if we parted company with one and not with the other, what we were going to do.

To his credit, even when the lawyers said the smartest thing to do or the safest thing to do—I don't know if they said the smartest, but they said the safest, because that's what lawyers do, that's what they're paid for—would be to part ways with all of them, to tell the FCC, "Look, we get it, we're with you; we've changed, we've fired these guys," Jeff was the one who stood up and said, "No, this is a free country. There's freedom of speech, and we're not just going to kowtow to a witch-hunt." I respect him for that until the end of time. Thinking about my radio career, that's a moment I'll never forget—Jeff saying, "No, we're not going to do that."

WOODY

In that time, I got some inquiries from the FCC, but all the stuff that got sent over from them was stuff that happened way before Janet Jackson. It was all this retroactive reaction. All of a sudden, there was an indecency crackdown; so then, they started going through their backlogs and trying to show that they were doing something about this.

JIM "JESUS" LYNAM

You couldn't talk about certain things; you couldn't use certain words. We were in such an unknown time, at that point, because you honestly didn't know what would get you in trouble. Could you say vagina now? No, probably not. What word could you use for vagina?

WOODY

I seem to have this talent for getting to stations at a very awkward time in their history. You know?

MIKE STERN

At one point, we even had posted a job listing—we never ended up filling this position, but we posted a job listing—looking for somebody who was going to come in every morning at 5:30, sit in an office locked away from Mancow, and ride an extra dump button. And anything that person found out of bounds, they were going to dump … That is the definition of the worst job ever.

We literally posted that job. I interviewed people for that gig, but we never ended up hiring (for) it. That had that much momentum coming from corporate, or however it got rolling; that had enough momentum that the job was posted—I got the resumes, interviews in place. I remember interviewing a guy who was like a former security guard or something, who wanted to move on from that. I thought, "Well all right, if he's tough enough to sit on his own and be a security guard maybe he could do this gig." But yeah, they were going to be locked in a separate room with a dump button. It was an ugly, ugly, moment in time.

WOODY

Besides the obvious seven dirty words, everything else—context, and community standards—who can tell? All you can do is do the best job without going out of your way to try to say something to try and shock people or disgust them.

CHUCK DUCOTY, *General Manager*

Justice Potter Stewart's famous quote out of that, when he was asked to define pornography, was "I can't tell you what it is, but I know it when I see it." I would say to you that the FCC's definition of indecency that they gave to radio stations was a paraphrase of that, which was, "We can't tell you what it is, but we know it when we hear it." When you've got any kind of aggressive morning show, that's a really difficult standard to try and figure out what it is; you're wrestling with a ghost.

WOODY, TONY, AND RAVEY

CHAPTER TWENTY-FOUR

One of the bigger on-air shakeups of the 00s came when heritage jock, Sludge, was bounced to mid-days from afternoons to make way for a trio of out-of-town talents making their first appearance in the Chicago market: Woody, Tony, and Ravey.

RAVEY, *Air personality*

I think it was a tough pill for (Sludge) to swallow when we came in, because he got bumped from that afternoon spot to mid-days ... I think he did his best to be cool with us. We were cool with him, and he did his best to be cool with us.

WOODY, *Air personality*

If anybody could have had a problem with us, it would have been Sludge, because he was the one who was doing our shift before we came in. But he didn't. I still consider Sludge a friend, and I reach out to him from time to time.

NATALIE DiPIETRO, *Director of Fun and Games*

Truly, (Sludge is) the funniest person I know. I miss seeing him every day. He and Ned Spindle together are just insane with ideas—the two most creative people on the planet.

DAN "BASS" LEVY, *Producer*

There was not a single bit of ego or anger to him. I've worked with Mancow, Chet Coppock, Mike North ... with Sludge; it was just, "How creative can you make this?"

MICHELLE RUTKOWSKI, *Air personality*

Sludge was the most fun to work remotes with. He always seemed genuinely to enjoy doing them and was the absolute best at interacting with people in the crowd. Oh, and if it was a daytime remote, he usually bought us lunch.

SLUDGE, *Air personality*

Mid-days was suffering after Mancow; there were a lot of people who only listened to Mancow and didn't listen to Q101, and they were trying to figure out a way to extend that as soon as Mancow got off the air. (Mike Stern) said, "You've always been tied with Mancow, 'cause you worked with him at Rock 103.5, and then you worked with him here for years. We just want to try an experiment, if you're up for it, to get you on the air in mid-days, and see if you can keep people around longer. Of course, the other end of it was Mike Stern wanted to bring in Woody, Tony, and Ravey, a new show, in afternoons. And I certainly wasn't failing in afternoons, but he wanted to try something different in afternoons. It's hard not to take offense when you feel like, "They don't like me. What's wrong?" (Stern) said, "It's not that at all. I just really want to try and shake things up in the line up a little bit, especially since the Zone is strong competition for Q101 at that time." Once I sat on it overnight, I still got angry. I was, like, "I don't want to change; I did afternoons for almost six years and I don't want to move, and I feel like I'm being demoted." But really, at the end of the day, I just had to think about it, and go, "No, I'll do it. I want to try it." Even though Mancow used to go over a lot, which took time out of your show, it did put you around that show, which had a lot of naked women walking in the hallways sometimes, so that was nice.

WOODY

Q101 was the station that was trying to figure out how to bring some buzz back to the station outside of Mancow, which obviously was its own standalone thing, a separate universe from the rest of the radio station. So, they brought in this afternoon show. I was the host of that show with Tony and Ravey, and they wanted us to be the afternoon answer to Mancow. I had a ratings history from the station I was just at in St. Louis, another Emmis station, and that's how we were brought in. We got there, and our contract was structured in a way that, if anywhere in the first year that we were there, we hit a certain rating … it was like our overarching goal of "Hey, if we can somehow accomplish this in the first year, man, we'd really be happy". Well, in the first ratings book we hit that, and in the second ratings book, we did even better than that. It guaranteed an extension on our contract. No sooner did that happen, (than) the station started fucking around with everything, and they started changing up the music. It went from being a male-targeted station, to all of a sudden, we're playing Death Cab for Cutie, and a lot more indie kind of alternative stuff, which is fine, but it seemed like every time they'd make an adjustment, it was to the polar opposite audience of what they were previously targeting. We'd go from targeting men, to 60/40 women; there was just no clear focus of who we were supposed to talk to. Meanwhile, Mancow's doing his thing. There was just no synergy; there was no common thread to the whole station. After two months of not getting where they wanted to go, they would just (have) a kneejerk reaction, and then go from playing indie rock back to playing nu metal, and then, from nu metal to really heavy doses of retro and grunge. From that, they went to the Shuffle thing. Every time that would happen, we would get told, "All right, now you're talking to this audience." There's no way to hold or build an audience when you're doing that. That was when we kind of knew that our time there was not going to be all that long. It was driving me crazy.

RAVEY

In the beginning, it was really, really hard because it was something brand new. Woody and Tony had worked together previously, so they had a rhythm. They knew each other's tendencies. They knew how to do a radio show together. They didn't necessarily know how to do a radio show with me in it, so it was really tough at first, just trying to figure out how I fit in to this dynamic.

SLUDGE

At first, I didn't want to like Woody, Tony, and Ravey, but I still took to Woody, Tony, and all of them a little bit.

RAVEY

Mike Stern asked me to take improv classes because, I don't know, maybe I wasn't funny enough for what was going on. So I went to some improv classes, I think at the ImprovOlympic. It was fine, and I enjoyed it, but it was working too hard. I think our philosophy was we were going to be as natural as possible. For Woody, that kind of meant, really, kind of skating the line of political correctness, and sometimes jumping off.

WOODY

The Program Director was trying to get the other two people on my show to take improv classes. Well, what does that have to do with anything? Tony, Ravey, and I were all legitimately close friends. I had known Ravey since I was 17 years old. Tony, I met when he was an intern at another station I worked at, and I ended up hiring him to be on my show and developed a groove from that. We knew how to interact with each other. Our show was never about bits, and skits, and song parodies and stuff like that; it was just people bullshitting, and breaking balls and talking shit about whatever was going on in the news, water cooler kind of stuff, media stuff, sports stuff, whatever. And that's just kind of, what worked—being interactive with the audience.

RAVEY

We got a lot of helpful suggestions, but not necessarily people telling us how to run our show. They let us run it how we felt it should run. We kind of got a vibe from the listeners and the community about what we should be doing. I actually thought and felt that the show got embraced right away, and it got really good vibes from the people of Chicago, right off the bat.

WOODY

We'd have these daily meetings with the Program Director, like, "Well, that could've been a little shorter." You know, it was, like, man, here's a person who's never been a successful on air person. It was like somebody who's never faced a major league pitch trying to tell Albert Pujols how to hit a baseball best. You might know programming, and (you might know) how you want to rotate your records and stuff like that, but as far as like being on the air, the personality side of things, it was just really frustrating. I was, like, "All right, we'll do this," and the next thing you know, we were playing six songs an hour. That became seven songs an hour, eight songs an hour. Then, we had a mandate that we had to play ten songs an hour.

CHUCK DUCOTY, *General Manager*

I really thought the radio station needed to build itself on more personalities. I don't necessarily think, in the long run (Woody, Tony, and Ravey) worked the way everybody wanted them to; it was still there when I left.

RAVEY

The story became imperfect for us when Chuck DuCoty left. He was a huge supporter of ours; he was the one who really had the vision for our show. He always had the best things to say, always encouraged us, always said if he liked something the day before; he was always very quick to tell us. I really, really thought we were on the right track. But then his life took him to a different place, and I don't think anything at Q101 was ever the same for us, once he left.

WOODY

On a personal level, I really liked Mike (Stern), but from a talent standpoint, I found it more frustrating because it was a lot of micromanaging, and I don't work well that way.

MIKE STERN, *Program Director*

We got a notice about a bit that Woody did; it was the dumbest thing ever, because it centered on the fact that he said the word "taint." Some listener was upset that he used the word taint; that was what we were going to get fined over.

WOODY

It was an inquiry. An inquiry wasn't necessarily a violation; they were looking into it to see context. They were investigating a complaint, and it turned out to be nothing. You could say, "Taint." What does taint even mean? Nobody even knew what that meant. It was no man's land, literally. That turned out to be nothing.

MIKE STERN

After that complaint came in, Emmis negotiated a consent decree, which is essentially a way of saying we paid the FCC a bunch of money without admitting guilt and the FCC said, "Okay," and wiped the slate clean. So, we never said we were guilty of anything, and they wiped the slate clean of all the old infractions and we paid them a bunch of dough for it. It was very much like traffic school. It was kind of like getting the ticket but having the points erased.

FOOK, *Air personality*

I liked Woody, Tony, and Ravey. I thought they did an interesting show. I thought Woody was a really talented guy. I never really hung out with them too much, so I don't know them personally.

They were trying to put on a morning show in the afternoon. They were all very talented dudes; it just wasn't working on the station at the time, primarily because the station wasn't working.

WOODY

They were paying three people to play ten songs an hour.

JACENT JACKSON, *Assistant Program Director/Music Director*

At the time, we had an airstaff that actively hated the radio station, which I thought was awesome.

You had Mancow on in the morning, who had never really hidden his contempt for the station on the air. Then, you had the afternoon show that wanted to do a four-song-an-hour afternoon show. I scheduled eight, and they would pick the four I didn't want them to pick.

WOODY

We had daily Uno games going on. Because we literally—this is really funny—they had mandated we played 10 songs an hour, but yet they still wanted content. I remember sitting in the Program Director's office. I was, like, "Hey man, I'm not trying to be an asshole, I'm trying to figure this out because I'm having a hard time with it; I'm having a hard time figuring out how this is going to work. You want 10 songs. You average out 3 1/2 -4 minutes per song." Plus the commercial loads were ridiculous at that point. So, if you added that together, it literally left—it was within single digits, the amount of content (time) for the hour. I was, like, "Where is that content coming from that you're looking for?"

RAVEY

The biggest surprise I ever got, artist-wise, at Q101 was an afternoon we spent on our show with Billy Corgan … he refused to go on with Mancow, so he was going to come in with us in the afternoon. We were looking at each other, like, "Billy Corgan, are you kidding? Oh, God, that's going to suck." I didn't think he presented himself all that well, necessarily. There was an elitist air about him and stuff; nobody was looking forward to this, believe me. He came in, (and) he was the nicest, coolest, most fun guy. We spent two awesome hours with him. For God's sake, he and Tony bonded for life, talking about wrestling. Billy Corgan was a wrestling freak ... I was stunned at how personable, how really nice, he was. We spent this great couple of hours with Billy Corgan.

RYAN MANNO, *Air personality*

(Woody, Tony, and Ravey were) hit or miss. I feel like their show would've been more successful had they not talked about St. Louis so much. They came into Chicago acting like St. Louis was better, which was (against) cardinal rule number one. I never really got a sense they were—I never trusted them for some reason. I always felt like they were ... that they almost didn't belong in Chicago and they knew it. They just seemed uncomfortable. There was something smarmy about that whole crew.

WOODY

Right when they started the whole "Shuffle" thing was when they started getting the idea that they were done with Mancow. I went to dinner with Marv and Mike Stern; they sat me down, and they said, "We have a plan for this radio station and that plan includes moving you guys to mornings." ... I was, like, "Well, when do you see this happening?" "It's probably going to be six months." Then six months came, and then six months went, and then, I was told another 90 days. "Ah, well, we're having a hard time; it's going to be another 90 days." Next thing you know, it had been over a year. And I was going, "This isn't happening." Now at this point, we were playing ten songs an hour (and) we were playing cards in the studio. I was just, like, "Howard Stern's going away. He's moving to satellite. There's going to be a lot of opportunity for shows. Let's just see what's out there." We had a countdown going on in our office, as far as when our contract was going to be up and we could start talking to people. We ended up getting let go from Q101, and from a business standpoint, it made sense for them to get rid of us, because they didn't need three people to do what they were asking us to do. I'm sure our attitude, from the outside, looking back at it ... we were pretty disenfranchised, I guess you could say.

SHERMAN, *Air personality*

Two thousand and five, that's when I went to nights. Fook went to afternoons because they fired Woody, Tony, and Ravey.

ELECTRA

CHAPTER TWENTY-FIVE

Electra, one of Q101's most revered personalities, was also one of the few who managed to stay in the same day-part for her entire time at the station.

Rock stations across America have long been populated by women who lean on "sexy" or "dumb girl" crutches—you know, the jocks who come out of a Foo Fighters song and say: "Oh my God, I'd totally do Dave Grohl!" Christine "Electra" Pawlak was a rare exception—an intelligent, educated, female voice who didn't play up her "femaleness" on the air.

Her arrival followed Sludge's amicable departure.

MARV NYREN, *General Manager*

(When I got there), it felt like there were a lot of people who were looking for direction—it wasn't about local management, from the standpoint of the format—the alternative format was going through a lot of stuff. There were months and quarters where product was great, and then you could go six months where nothing new was released. I think alternative was trying to find itself and what the next level for alternative (was).

SLUDGE, *Air personality*

There was a changeover in management with GMs and PDs, and no one was paying attention to me. My contract ran out without anyone knowing about it. In reality, that was the only way I could leave. I remember when I told Mike Stern about it, I was pretty honest and upfront. I went, "I just want to be honest with you. I got a job offer from the Zone to do mornings." He was, like, "You can't leave." I went, "Actually, I can. My contract's over." He went, "When did that happen?" I went, "About three months ago."

They let me stay on the air while I was making my decision for three weeks. That was unheard of. Usually, they'd have at least pulled me off the air right away or just said, "Go. Fuck you. Get out of here." Marv came in at one point, made an amazing offer to stay … (which) made it much more difficult to leave; it was a long-term deal to stay, and really good money. It was one of those moments when you go, "That's awesome, man, people do care."

CHRISTINE "ELECTRA" PAWLAK, *Air personality*

Someone said, "Hey, you should send your resume to my friend Mike Stern in Chicago." I think it was March, when Sludge had left. So I sent my resume and my demo. At the time, I thought Q101 was not a good fit for me because it was harder rock. I was 24 and a very alternative-angsty chick, but I needed a job so I sent my demo and my resume and I didn't hear from Mike until May, which was not all that long, considering that there are people who are out of work in radio for months and years. But at the time, since it was the first time I had been fired and the first time I was trying to get a job, it was agonizing. I talked to Mike on the phone; I talked to Marv Nyren on the phone; they wanted to offer me a job over the phone. I said, "Can I at least come out and see Chicago first?" They flew me out. I saw Chicago, and I took the job.

POGO, *Air personality*

I had known of her prior because she got a job that I wanted in Philly when I was in D.C.: nights at Y100.

MIKE STERN, *Program Director*

(Hiring Electra) is something I'm very proud of. She applied; I think I posted the job and she sent me the standard resume and demo. The Philly alternative station, Y100/PLY, had either flipped or was about to flip. Mike Parrish at FMQB, which is based in Philly, said, "You really oughta take a look at this girl at Y100 here, Electra." Her tape stood out in a good way. I narrowed it down to a couple of people, and she was clearly the one to hire. Female air talent is hard to come by, and she clearly stood out.

CHRISTINE "ELECTRA" PAWLAK

When I got to Q101, I knew there were a lot of people who had just come to Emmis, with the Loop being acquired—a lot of people from Arizona being freshly planted in Chicago. There were a lot of people who were all trying to get their footing in establishing themselves at the station. So what I felt like ... there were a lot of defensive people. There were a lot of people who were adjusting to change and trying to figure out what their roles were, so I encountered some prickly people when I first got there.

MIKE STERN

The vision (for Q101) was a little smarter, a little less pandering, and Fook fit right in. Electra was a perfect piece for that, and Fook fit right in behind her.

CHRISTINE "ELECTRA" PAWLAK

It wasn't until after I accepted the job that Mike Stern said, "So, about your name …." And I said, "I've been Electra my entire radio career so far; I didn't know my name was going to be an issue." "Eh, moving forward, thinking about branding, maybe you want to think about being Christine…." I was sort of taken off guard by that, especially because it hadn't been broached until I already said yes. I fought him on it, and we had an on-air contest; that was Mike's compromise, where we had listeners text in Christine or Electra, which automatically made me feel a little uncomfortable.

MIKE STERN

I always thought of her as Tina Fey as an air talent; she was a smart, snarky, pop culture-savvy, air talent. If Tina Fey had decided to go into radio, I think she'd have sounded like Electra did.

CHRISTINE "ELECTRA" PAWLAK

The person who I first met after being hired was Alex Quigley, who immediately let me know that his good friend Sherman had wanted the mid-day job that I had gotten, and I should be aware of that. There was a lot of bluster, but over the course of four hours, Alex realized that I was not a bitch, that I liked Final Fantasy, that I could talk about sports, and that I wasn't there to stab anybody in the throat and take over the world. We ended up at Big City Tap. The first and only time I've been to Big City Tap in my life was after my first board op training with Alex Quigley, the first night I got here.

AJ COX RUDLOPH, *Sales Promotion Manager*
Electra was pretty much the most solid show on the station for the six years she was on.

TISA LASORTE, *Brand Manager*
She was the stable core of Q101 during my time there. I'd say she kept the station afloat.

CHRISTINE "ELECTRA" PAWLAK
Mancow wouldn't always get off the air at 10. Sometimes he'd get off the air at 10:15, 10:30, 10:45 — whatever. Cow time was Cow time. Within the first few weeks that I started working at Q, I was in our jock lounge at the time; it was after 10. I was doing show prep, or just browsing the web or something, and two of our promotions assistants were in there with me, on the other side of the room, (with their) backs to me: Kevin Manno and Gordon Mays. We had the station on, because I had to listen to find out when Cow was signing off to know when I could go in the studio. I hear Cow say, "All right, Stella's up next." And one of his crew went, "Cow, her name's Electra." And Cow said, "Doesn't matter — she won't be here in six months anyway." I remember that my eyes kind of welled up; I am a little bit emotional. I definitely felt like the new girl. I've been the new girl a few times in my career already, but never like that, where someone publicly, whether it was part of his on-air persona or not, called me out to fail. And later, Kevin and Gordon both said to me that they felt so bad for me, that they didn't know what to say — what could you say? Nothing.

But … Cow and I do share a love of the Kinks. The man's not all bad.

TIM VIRGIN, *Air personality*
She was awesome, and to this day, I believe that she is the greatest alternative female disc jockey – I think she's the greatest female jock that I've heard, ever, at this point. For the Q101 brand of listeners, she was like the G4 of chicks. I think she gets that lifestyle more than anybody, which makes her the best jock of this format.

KEVIN MANNO, *Air personality*

Everybody loved Electra; she was "Ms. Q101" for so many years there at the end.

CHRISTINE "ELECTRA" PAWLAK

Chicago was very welcoming to me. I grew up in Rhode Island, and I had never even dreamed of living in Chicago. And I still remember being new here, and calling it "Soldier's Field" and having a listener call up and say, "First of all, welcome to Q101. I think you're doing a great job. Secondly, it's 'Soldier Field.' Don't worry about it; people in the suburbs call it that, too, so you're going to be fine." Listeners did not make me feel unwelcome. I don't like using double negatives … listeners made me feel welcome. There, now I feel better.

JEN JAMESON, *Air personality*

Besides Electra, I don't think us women have really been given that great of a shot.

CHRISTINE "ELECTRA" PAWLAK

The funny thing is, despite fighting to be Electra on the air, it's not a persona; it's just about as close to who I am as I feel comfortable being on the air. In some cases, even amplified. I do enjoy sports … I don't like being defined by my gender. I wanted people to enjoy listening to my show and think that I was a good DJ, a comfortable sweatshirt that you huddled with on a cold Chicago afternoon—something constant and reliant and dependable, but not female, per se.

FOOK, *Air personality*

Probably the best natural host I've ever heard. She's a true natural talent.

CHRISTINE "ELECTRA" PAWLAK

Oh, Fook. I was fascinated by Fook because he seemed impervious to my charms. He was an angry, angry, man and so I thought, "I'm going to make him like me!" I tried, goshdarnit.

KYLE GUDERIAN, *Operations Manager*

Electra. To me, (she is) probably one of the, if not the, best female air personalities in rock radio today, and probably in radio in general.

TIM VIRGIN

She knows so much about the music; loves the music … she's an A-lister, no doubt.

KEVIN MANNO

She just became so comfortable in that timeslot … that was the most natural, perfect fit for her.

CHRISTINE "ELECTRA" PAWLAK

I never wanted to come out and say, "Look at me! I like chicken wings, and I like beer, and I like rock music!" I just sort of let what I was passionate about show people who I was, and that sort of thing takes time. Building that kind of relationship takes time, but I didn't want to walk into Q101 and be like, "I've got tits, look at me. Or don't, because it's radio, and not a visual medium."

RYAN MANNO

I really think Electra was probably the best at what she did. She found her audience quickly and she knew exactly how to speak with them every day. She was so good at saying exactly what she wanted to say in as few words as possible. She's so endearing.

SPIKE, *Assistant Program Director/Music Director*

Electra was like my first friend when I got there … As a friend and as a midday DJ who worked for me, both ways, I could always count on her.

RYAN MANNO

I think she'll go down as a definitive voice of Q101.

ALL MIXED UP: Q101 "ON SHUFFLE"

CHAPTER TWENTY-SIX

POGO, *Air personality*

After the Zone switched formats to cock rock, I was starting to look across the river at the Merchandise Mart, like (makes puppy whimpering noises). It was folks like Nikki who would see me out at cool shows all the time, the indie rock shows, and things like that, and she'd be, like, "Why don't you work for us? What are you doing over there?" She kind of waved my flag for me.

All of a sudden, one day out of nowhere, Jacent contacted me, kinda like, "Do you want out of there?" I was, like, "YES." At that point, I was producing the morning show; I was hating life.

JACENT JACKSON, *Assistant Program Director/Music Director*

There was a borderline obsession with the Zone as the competition in the market. And the research we were getting back was showing that the stuff that the Zone was playing was the stuff that the market wanted, which was why both radio stations were two tracks deep on Slipknot.

MIKE STERN, *Program Director*

So we were pretty heavy and hard, just like they were, because we had Mancow. So, we had all those guys coming to us, and if we were not playing the right music, they were just leaving.

BILL GAMBLE, *Program Director: 94.7 the Zone*

You ended up with two radio stations, both about a 1.7. For the life of me, I mean, at some point in time, I guess I'd have to see the other side, but it was—there was a business for us and a business for them. But, you know, I don't know whether it was Mike Stern or what, but I gotta tell you, I still to this day don't get it. 'Cause there was room for a poppy alternative to push Norm (Winer, WXRT Program Director) and also to push TMX a little bit, and give Q101 every 18 to 49 year old guy in the market, so they could just push The Loop and be the rock station with Mancow. And, it was not just me. All the close, really great programming minds at ABC, they all thought the same thing! Because we played out all the different scenarios: "Okay, well, they can't block you; they've got this wide open rock lane." And then, when we finally decided, "Well, they're going to be alternative. We'll come and be the rock station." "Well, they won't follow you because they've so invested in alternative." "Yep! Here they go!" It was just like, "Okay, really..." "It's like a kamikaze mission here. So..."

JACENT JACKSON

I was playing "Before I Forget" by Slipknot on a two-hour and five-minute rotation, and was just like, "What are we doing?"

MIKE STERN

So, we got in the heavy and hard battle up against the Zone. I'll admit that was my predisposition; I started Extreme Radio in Vegas. That's where I came from; so, that was natural for me.

JACENT JACKSON

There was a period where the Zone was literally beating Q101 outright. It happened in a lot of monthlies; there was a point where Q, I think, was at a 1.4 or something, and the Zone was at a 1.8 or some crazy number like that. Neither was great, but the fact that you've got a three-million-dollar morning show on the air, and you're doing a little over a one-share in the market, and you're not winning your demos, that was just completely demoralizing.

MIKE STERN

It was a tough spot—the station wasn't doing well. The Zone had serious momentum. They were up in our grill.

BILL GAMBLE

We said, "No one on the planet's gonna rock any harder than we do." And no one did. It was just … The Zone just … it made Rock 103.5 look like a little girl radio station. And then, oh jeez, look at The Mix. They're like a … I don't know what are … they … a 97 share now in the market?

POGO

Nothing against the Zone, but ABC's not the same sort of company you'd associate with a rock station.

MIKE STERN

After a while, we realized that neither station was getting anywhere. We were trading a tenth of a point every trend, and it really didn't amount to anything. I was on the phone with the guys who were consulting us at the time; I was on the phone with (consultant) Dave Beasing. I was, like, "We've gotta do something different," and Dave said, "All right, let's tighten it down." And I said, "You know what Dave? The whole world is going the other way," because JACK-variety stations were popping up. Like, why is the whole world going wider when we're going to go tighter? And that got us talking about creating what became "On Shuffle."

JACENT JACKSON

The second year I was there was, "Well, we're never going to be able to out-Damage Plan these guys."

MIKE STERN

Knocking our head against the Zone day after day wasn't accomplishing anything, and it was hurting sales. Our demos were getting so narrow; we couldn't sell it. Broadening the sound of the radio station was an opportunity to make more money, and ideally, to bring in more listeners as well.

CHUCK DUCOTY, *General Manager*

You just took one radio station that … there was probably enough audience to really succeed, and just carved that up to the point where, what was the purpose?

MIKE STERN

The one thing I would probably do over, in retrospect, is I probably would not have spent as long, or even at all, being as head-to-head with the Zone as we were. Playing Slipknot and Mudvayne 30 times a week didn't help anybody. In retrospect, a more mainstream approach based on the station's heritage probably would have made everything a lot better. I didn't see that at the time.

BILL GAMBLE

Yeah, thanks, Mike. —Bill Gamble, 2011.

Once Q101's programming elite had realized that the station had veered too far in the hard, or "active," rock direction, they worked at finding their footing again. The solution and recovery plan was an "anything goes" format named "Shuffle," which allowed Q101 to play songs from across the full history of alternative music.

"Shuffle" started as a one-off, weekend promotion.

JACENT JACKSON

That whole first weekend before "Shuffle," I was asked by Mike to put together a "Jack" version of an alternative station for the weekend, which was, basically, what Shuffle was. How could I forget the Jack radio jukebox? Remember that trend? Good times.

MIKE STERN

We tried it out for a weekend; there was an "On Shuffle weekend" before it became the full-on format; it was kind of an experiment. The phones did not stop ringing all weekend. I remember calling J Love at like one in the morning, and he was, like, "Boss, I've got all five lines lit." People were just blown away. We did get a lot of attention for it. We got a lot of buzz, because it was suddenly broad. From the time I took over to the time we went on shuffle, was the time we were head to head with the Zone for a stretch.

JACENT JACKSON

Mike was like, "Hey, we're just going to do this; this is going to be the format next week." I'm not kidding; like, "This is going to be the format … Next. Week."

MIKE STERN

"On Shuffle" was actually based solely on familiarity. The concept of "On Shuffle" was a lot of width—so, musical styles that ranged from Depeche Mode and the Cure to Mudvayne, as long as the research said it was familiar. That was actually the rule. "On Shuffle" was kind of like the concept of Jack FM; the original concept, not what it (has) become these days. The original concept of Jack FM was pop songs, dance songs, rock songs, whatever, from mostly the 70s, 80s, and 90s, just so long as they were familiar. That was how they had that "Oh wow" factor of "Wow, I haven't heard that in a long time." Because everybody knows Mister Mister "Kyrie," but nobody plays it anymore, because it doesn't test great, but it tests familiar. So the idea of "On Shuffle" was to be the width and breadth of alternative.

JACENT JACKSON

You can't fall off the floor, and I think I said that to one of the record reps the night we went "on shuffle." Shuffle was a complete and utter desperation move. That was a Hail Mary pass, and here's why. We sat in a meeting with Coleman (Research). Coleman did a lot of research for the radio station. They did this really extensive market study and pointed out different music genres, different ways to put them together, their format coalition builder, and all these things. What it really came down to, they would have these suggestions for what you should do with your station at the end of it. So we're sitting in Maggiano's downtown; we're in one of the banquet rooms there. They've got the power point up, and they're going through this. At the end of the presentation, they said, "Punt." That was the end of the presentation. You could have heard a pin drop in the room.

It was, like, "You're not winning any of your music images, your morning show is collapsing, and the appeal of the morning show doesn't appeal to the fans of the music, so basically you're in a box: punt."

It wasn't directly after that when it happened, but it kind of just rattled around in various folks' heads for a while. I can't remember exactly how we did the weekend, but we did the weekend. We did Shuffle for a weekend before we actually did it. And when we did it, it wasn't called On Shuffle, but it was, basically, these weekend things I used to do in Boise. I used to do these throwback weekends in Boise where I would grab a whole bunch of crazy ass, history of alternative, kind of unfamiliar stuff, and just train-wreck it. We didn't have anything else going on at the time, and I don't remember how we decided to do it; so, we train-wrecked all this stuff together. It was the only thing we did in the time I was there—it was a legit response, and the amount of calls and the amount of email—there was a really tall stack of printed out email in my office about how amazing this was.

AJ COX RUDLOPH, *Sales Promotion Manager*
I think it did create some buzz; I think it created buzz in the industry.

MIKE STERN
Going on Shuffle, we did for Q101, but I believe to this day, when we went on shuffle and we broadened out the music, and we stopped being head-to-head with the Zone, it wasn't long after that the Zone flipped. We got a ratings bump out of going on shuffle—not a huge one, but we got one. I still think, to this day, that part of what led to the Zone's flipping format was that the format was on its last legs; we had intel from inside the building that ABC was kind of tired of not being successful with that frequency. When we weren't suddenly head-to-head with them anymore, and the ratings didn't go up, it was kind of that last, "Well, wait a minute. Why are we doing this?" kind of straw. I think that played a role. It was kind of like,"You can't point to the bad guy anymore. The boogey man's not head-to-head with you anymore. Why aren't the ratings going up at the Zone?" And they didn't go up, because there just weren't enough guys in the demo that wanted to knock their heads against the wall repeatedly to drive it.

JACENT JACKSON
Here's how (Shuffle) was researched: for whatever reason, all the old music tests had been left at the station. I literally had all the old music tests from 1994 all the way up to present. I had virtually every music test Q101 had ever done. I took them home, and drank, and went through them. And then, I built some clocks; and then, I brought them into Mike's office; and then, we sat in Mike's office until really, really, late into the night, trying to figure out—walking through this stuff, and going, "Well, Peter Murphy 'Cuts You Up,' can we play that?" "Yeah, I guess so; it was number 202 out of 400 in this test in '96. Okay, sure." That was literally how we did it. Some of the stuff like (Dead Kennedys) "Holiday in Cambodia" was obviously never tested; that was just shit I liked to play.

The mix itself was 1000 deep when we put it on. That actually got a lot of traction. That was legit, but it was like nothing else the station had ever done.

SLUDGE, *Air personality*

I think the Shuffle was a disaster. I thought that was the dumbest idea ever. I wasn't there; I had just left for the Zone. I think that was the beginning of the misdirection of what Q101 was in the 90s and up until the early 2000s. I'm sure the people who came up with the Shuffle would agree now that it was a dumb idea. Keeping what Q101 always was, a cutting edge alternative rock station, and how it took care of the listeners got confused, I think, over the last several years, with different people running it, different program directors coming in from other cities.

MARV NYREN, *General Manager*

I think there was some positive—we (had) a little bit more mass appeal. We stole the Shuffle, obviously, from Apple, to put it bluntly; I hope I don't get in trouble for that one. But it was certainly a word we kind of took from Apple, from the standpoint of "we're going to play a lot more of what you like". So it gave us a little more breadth, width, of music. I do think it was well received from the advertising community, especially when I'd say 60-70% of the buying community was made up of females between the ages of 25 and 40. They were much more open to listening to Q101 in a shuffle format than they were to a harder, more male skewing station. From a financial standpoint, we did well; we continued to do okay.

SHERMAN, *Air personality*

It was murder on the air, at least for me personally, because you'd be talking out of "Fuel" by Metallica, and then you'd be going into Death Cab for Cutie, or a song that you needed a shotgun to come with the album. Talk about weird transitions …

JACENT JACKSON

I threw in things like Liz Phair, Smoking Popes, and Material Issue as a love letter to the old music scene. Things like Dead Kennedys, Fugazi, and Camper Van Beethoven were meant to be calling card records so people would know we were doing something different.

POGO

I'll hand it to them for going for it at a time when it was really shifting towards people's own personal content and things like that. That was the beginning of that.

KEVIN MANNO, *Air personality*

I feel like once you lose sight of who you are, or change it so drastically, you're making the listeners question it. They just don't have as much faith in you when you make such a weird, bold, change. I feel like it just lost something, at that point, (that) it never got back.

JACENT JACKSON

It started out based on old tests the station had done and gut checking a lot of it; and then, we decided that we should properly research it. How do you properly research that? How do you screen that? Who's in the room? What are their other choices? ... I think that the screener ended up being "they have to say they like alternative music." The station was playing like 1000 records; you can't test 1000 records ... what came back was ... I don't know ... it looked like we were playing "Take on Me" by a-ha.

POGO

For me, it was great. I understand the economics of it, and the formatics of it, and the radio business side of it, and why it didn't work, but for me, I couldn't wait to get to work. I couldn't wait to look at the log—sorry, people, yes even with the Shuffle there was a carefully planned playlist. I couldn't wait to look through the hours in my shift and see what was coming up.

BRETT "SPIKE" ESKIN, *Assistant Program Director/Music Director*

As a youngish programmer—I guess I was 29 at the time and I was coming from active rock radio—it was welcome as a programmer, because I wasn't used to programming things that sounded like train-wrecks or clashes. For me, who has a very eclectic music taste—I really really love Slipknot and I really, really love Metallica, but I also really, really like Dashboard, and I really, really like Depeche Mode— all those things in, like, this kind of mish mosh, when it was new to me, was fun. You know? It was really hard to define what it should sound like. I don't even think Mike Stern totally knew what it should sound like, whether we were a soft radio station, or whether we were a male or female (station), or whatever it was, it was just supposed to sound different.

As it went on—that kind of radio station, and you hear it on a JACK radio station— those train-wrecks, or those "wow" songs that you program in there, are only wow and are only train-wrecks the first ten times you hear them. Once you've heard that happen all the time, and once you've heard "The Safety Dance," like five times, then you just realize you were sick of that song ten years ago, and it goes from being a song that you haven't heard in a while to a song you heard yesterday that you're already tired of. The problem with Shuffle, and the problem with advancing it, was what you had to do was, like, either get wider, (and) you're already too wide in the first place, or go deeper and get these songs that people knew even less. So, you're not even getting these old hit songs that you haven't heard in a while; now, it's like an old album track and you're just trying to find different old songs to mix it up and you're playing worse songs just to make it sound like Shuffle. It felt like it was a good way to say, "Hey, we're different," and a good way to say, "Hey, we care about music," and a good way to address the listeners' concerns of hearing Beastie Boys and Slipknot every two minutes, but it had a limit. It had an end of the line.

WZZN, "The Zone," had struggled since it first launched in 2001. After years of slumming it out in the ratings with Q101, the Zone blinked first and flipped the station to a much safer and cost efficient "oldies" format (sample artists: The Monkees, Tommy James).

MARV NYREN

On one level, the excitement, when we heard that the Zone was going away, we were euphoric. We were still in what we called the old diary system in the Arbitron world, the ratings world. We still thought when the Zone left, there was no other station that was close to the music the two stations played; so, we really thought, this is it, we're really going to zoom back to the top. I was very disappointed, from a ratings standpoint, I really was. I thought we were going to get more of their audience to come on over, and while we got little bumps here and there, there really wasn't the positive impact that I thought (there) was going to be.

JACENT JACKSON

For me, it was, like, "Well, fuck, I'm glad that's over."

MICHELLE RUTKOWSKI, *Air personality*

Since I was only a lowly part-timer, I stayed out of the programming department's way. My discussions with Jacent regarding programming philosophy were limited to me asking, "Why is there a Goo Goo Dolls plaque on your wall?" and writing, "Please don't ever make me play The Tubes' 'She's a Beauty' ever again" on one of my music logs. But I guess, if you look at what he'd done prior in Boise, and what he's done afterwards in Milwaukee, and you couple those things with what happened to Q101 after he left, you could probably deduce that if he was part of the problem, it was a very small part.

POGO

I remember we unceremoniously let go of Jacent, who knew music, knew the format inside and out, had lived it his whole life and listened to it his whole life. All of a sudden, (Mike Stern) brings another rock guy (Spike).

He walked in the first day of his job, or something like that, to be the Music Director at the second biggest alternative station in the country, and he was, like, "I'll be the music director of an alternative station in a major market, and I've never actually even heard a Depeche Mode song in my life." I think James told me that story. And I was, like, "That's it," from day one; I have zero respect for this guy as a programmer of the alternative music format.

BRET "SPIKE" ESKIN

It was weird when I got there. When I got there, Mancow was still there, but I was aware that there was a very strong possibility that he was not going to be there. It was weird; it was, like, this crazy, aggressive, male morning radio show and the shuffle thing going on the rest of the day, which was pretty much anti-that.

As far as the programming and the music, it was weird, and everybody kind of knew Mancow wasn't going to be there. But as far as the culture itself, it was perfect for me. It was fun, it felt…there were mostly young people there; it was mostly people that enjoyed doing what they were doing. I came from a very corporate atmosphere at CBS.

When I came to Q101 and Emmis, it felt younger and more fun, and it felt like everybody, at least at that time, felt like there was opportunity ahead. There was a changing of the guard and a changing of the tone of the radio station back to a music-centric kind of image. It was a good time for change in me and in my life, and it felt like the radio station was about to go through a change as well.

THE COW IS NOW ... GONE

CHAPTER TWENTY-SEVEN

RYAN MANNO, *Air personality*

Jim (Lynam) would come in and do traffic on Mancow's show, and he would dip every morning. He would put an entire can in his bottom lip, and he would spit in these Red Bull cans that we had.

JIM "JESUS" LYNAM, *Producer (Mancow's Morning Madhouse)*

When I did traffic, I used to dip. Most of the time, I'd get called into the studio out of the office to do traffic. I'd try to get it out of my mouth or whatever, and I wouldn't have time, and I'd be like, "What the hell, I can do it with it in my mouth." So, I had been using a can the entire day, and I think it was in the 9 o'clock hour, but it was a full can of my dip spit.

RYAN MANNO

(Jim) had just done a traffic report. I had a Red Bull, but I put my Red Bull in the Production Studio where I was recording something with Abe (Kanan). Jim left his in front of the microphone I would share. I ran in the room and I slammed what I thought was my Red Bull. It was a can of four- or five-day-old Jim Lyman dip spit that was not just disgusting, but hot. I just threw that down the gullet and it all came right back up on the air.

JIM "JESUS" LYNAM

I finish my traffic report and I go into the office. I'm listening to the show, and it's in the middle of a commercial, and Mancow breaks through the commercial. He's laughing. He goes, "Ryan, what happened?" He goes, "I just drank all of Jim Jesus' spit." I walked down there, and Mancow's like, "What are you doing? Why are you drinking his spit?"

He goes, "I didn't know he had a can in here," and Mancow goes, "Oh, you just drink whatever's around?"

RYAN MANNO

I puked live on the radio. I was confident (the can) was mine.

AJ COX RUDLOPH, *Sales Promotion Manager*

We always had morning show broadcasts for Cubs Opening Day. We were at a bar in Wrigleyville; Mancow would broadcast. Those typically meant I was at work at 12:30 in the morning, and worked until about 2 or 3 o'clock in the afternoon. We would get up, set up the broadcast, make sure everything was good, sponsors were happy, blah, blah, blah. Cow's crew would come and get set up and do their thing. Since these were bar events, they were before a baseball game, and they were usually sponsored by a beer, the morning shows would generally proceed to get hammered during the shows and afterwards as well. And then, usually, they made their way back to the station mid afternoon or so, just completely annihilated. I had gone back to the station, finished up some work after the event, and had made my way home right before Turd wandered back to the station. I didn't actually see this. However, apparently he came over to my desk and our graphic designer Sabrina was sitting there; she sat at the desk next to me. He came over, and he threatened to poop in my desk. He pulled my top desk drawer open, and he said, "I'm going to poop in her desk." I have no idea why; he just wanted to poop in my desk. He did not poop in my desk. Apparently, after that he walked over and threatened to poop in the trash can next to somebody else's desk … then proceeded to walk down the hallway and take all the picture frames off the wall … took all of those down, put them in the Program Director, Mike Stern's, office.

The next morning, Cow talked about all their shenanigans on the air and made them all apologize to everybody that they messed with. So I got this adorable voice mail on my phone the next morning when I came in. It said, "AJ, it's Turd. Sorry I tried to poop in your desk. Love ya."

Kinda wish I'd have saved that voice mail, because it was the best ever.

JUSTIN STEPHENSON, *Lead Application Developer (Emmis Interactive)*

I always thought it was interesting doing this tech job; we were in the back office, basically, like, stuck in a closet behind Q101. So, we had to run the gauntlet past Turd and Mancow. Then, once upon a time, something happened where, I think, we were moving our office from one place to another. We ended up working from home for a while, and on the first day back in the new office, we came in and there was a whiteboard, a typical, corporate whiteboard, and written on the board there to welcome us was "Welcome back, nerds." Not that we could really argue that, but …

MICHELLE RUTKOWSKI

With a few exceptions, (Mancow's crew) was a bunch of fucking bullies. But they could be, because they got away with it. It was their *modus operandi*, and everyone just kind of accepted it because they were told it was for "entertainment value."

MARV NYREN, *General Manager*

I think when we were looking at costs, and we looked, specifically, at Mancow's morning show. We had approximately ten employees on the morning show. One of them made a lot of money. Two of them made a pretty darn good living in any market, and then there were six or seven other people making decent money as well. It was a very expensive morning show.

MIKE STERN, *Program Director*

You have a time when budgets are tight, marketing money's gone, budgets are getting cut, things were not great financially for the company, and here you have a guy who's expensive. And not only is he in an expensive contract, but with him come five or six other expensive contracts—not at the same level as him, but that entourage wasn't cheap. Collectively, it's a lot of money.

MARV NYREN

It generated good ratings, but (it) also generated really good ratings in demos that were harder to sell. There weren't as many buys up for men 18-34 as there were years ago. We really had to focus on beer money, younger-skewing automotive business, concert business; it became a much smaller group of accounts that wanted to spend money.

AL ROKER, JR., *Air personality*

There had been rumors that they were looking to replace us … they had acquired The Loop at that point. Brandmeier was working; they'd moved us out of the studio they built for us, into another studio, put Brandmeier in our studio. Then, (they) put up a carwash divider, put up the carwash things, in the hallway to separate us. So, it was not a great atmosphere.

MARV NYREN

I believe that Mancow was an extremely talented guy and did a wonderful job of building a brand and growing it. I'm going to paraphrase a little bit, but when I sat down with him a month or two before we made the final decision, we started talking about what we could do differently to make this a little more friendly radio station for advertisers, for ratings purposes. His line was something to the effect of, "This is the girl I brought to the dance; this is the girl I'm going to leave the dance with," meaning, "I'm not changing." I heard that repeatedly, and I think that was the other factor that said, not only does the show cost us a lot of money, but also he's not going to change it, and therefore, we have to make a decision.

The decision not to renew Mancow's contract was made public in July 2006. Station management informed the "Madhouse" crew.

RYAN MANNO

Goddamn, that was a crazy day. We all went into Marv's office after the show, and we sat in a semi circle, almost like we were in Kindergarten. The entire crew of the show, minus Mancow, got the news from Marv that the show had not been picked up. So what they did was let everybody go so that no one felt weird or slighted, and there was a stretch during that afternoon I thought I was out of a job. They made no mention of "some of you we're going to talk to about staying;" they just said, "Sorry, that's it." So, we all went over to Shamrocks and I got a call from Cliff Mazzone (Market Controller), and he said, "Hey, come back over; we want to talk to you." I found out that me and J Love and Jim Lynam—they asked the three of us to stay but they didn't want to say that in front of everyone else.

TIM RICHARDS, *Program Director*

Mancow and Q101 should've parted company long before they did, and that would've probably helped the station's long-term prognosis. It's just an opinion, but I do believe that.

JIM "JESUS" LYNAM

It happened so sudden, because Mancow didn't really tell us anything; he just kinda left.

As everybody's walking out, Marv goes, "Jim, can you stay for a second?" I said, "Sure." He said, "I told Mancow that I want to keep you no matter where he goes."

MIDGE "DJ LUVCHEEZ" RIPOLI, *Technical Producer (Mancow's Morning Madhouse)*

(We learned that) everybody was fired except for Jim Lynam, who was going to be staying on. We were all just, like, "Oh, you fag. Thanks Jimmy."

MIKE STERN

The question was, "What are you going to do to really … if you're going to do this, what are you going to do in morning drive?" Following in the formula of broader, smarter, and not pandering, the idea that became the Morning Fix started to germinate.

BOULEVARD OF BROKEN DREAMS: THE MORNING FIX

CHAPTER TWENTY-EIGHT

Once Mancow was gone, Q101 launched a new concept in morning drive: an ensemble-driven comedy show designed to be the FM version of Comedy Central's Daily Show.

ALAN COX, *Air personality*

My agent called me and said, "Hey listen. Mike Stern at Q101 is putting together something interesting—they want to do a Daily Show/SNL type thing. They really want to try something different because they're going to get rid of Mancow. I asked him, "Is this a bluffing tactic by Q101 management to get Mancow—are they just saying they're going to do all this stuff?" He said, "No, he's out, and they want to put this ensemble show together." I had done a solo show my whole career, but I wanted so much to come back home to Chicago. And for it to be at Q101, it was just unbelievable kismet.

They said the host of this thing had to be somebody who knew comedy and radio, because I had done standup for years before I went all in on radio, full time. So I was really the only one of the group who had a background in radio and standup. I went in for a couple of auditions ... and they hired me.

MIKE STERN, *Program Director*

I had known of Alan and our consultant Dave Beasing for a while; (Dave) was probably the co-founder of the show along with me. This idea really became Dave's and my project. Dave knew Alan, and Alan had that type of sensibility. He was smart; he was on the air; and he wasn't doing dick jokes ... he had the right head for it. He was the guy; he could do it. And there wasn't a lot of talent out there that really fit that bill. Most radio talent out there were doing dick jokes and pandering, and didn't understand anything different.

ALAN COX

When they sat down to explain the concept to me, it all seemed like it could work, theoretically. I knew that practically, it was doomed. But I didn't care, because I was going to be at Q101 for two years, and I figured, by that time, hopefully, I would have established myself at home to some degree, to where maybe that would be my last stop. That's not the way it worked out. I would be lying if I said that I was, like, "This is just going to blow the roof off the joint." But it was a great group of people, and it was fun while it lasted. We were replacing Mancow, for crissakes. You don't want to be the show that replaces an extremely popular, high profile, show.

MARV NYREN, *General Manager*

Loved the concept of *the Morning Fix*; I loved the people that were brought in. I thought that some of the things that were created were wonderful; they were just on the wrong radio station. It was an older-skewing morning show for a younger-skewing radio station.

MIKE STERN

It was an attempt at an attitude and intelligence similar to the *Daily Show*, taken to morning drive. It was also an attempt at a different set of formatics from the typical morning show, in preparation for PPM (Arbitron's audience measurement system). The other thing that was part of it was (that) when you talk about the sensibilities and humor of the *Daily Show*, you talk about writing. Radio stations, once upon a time, had writers, and they didn't anymore. The idea was to have people who could write prepare the show in the off time, and be topical and timely, and have stuff written ahead of time.

STEVE TINGLE, *Air personality*

I had no idea what the show, what the content, was when *the Morning Fix* started. I thought it was going to be a regular type show. When I sat down with

Mike Stern, he told me what it was. I was, like, "Whoa, this is different, Never heard of anything like this before." I didn't know what kind of role it was going to be. I was willing to do whatever; I was just glad to be out of a rock station. I wanted to get back into an alternative station and come to be me, and then incorporate my voices into my kind of show and what I do.

JIM "JESUS" LYNAM, *Air personality*

I remember when they told us, because Tingle had moved his life from Arizona to here without really knowing—"We're going to do a new kind of show" is what they told us. I had signed on. I had been doing sports for James VanOsdol, and when they brought us in collectively and told us, we were, like, "I don't get it. You're not going to take phone calls? You're going to repeat the same report?" I don't get it. Me and Tingle walked out. We went into Shamrocks, and we sat down and looked at each other. We were, like, "What the fuck is this? I don't even know what it's supposed to do. Is it a news show? Is it funny? I don't even know."

STEVE TINGLE

As soon as we knew what the show was about, me and Jim Lynam went across the street to the Shamrock and we sat down next to each other. We ordered a beer; we each got our beer in hand, and we just turned and looked at each other, and we went, "Uh-oh." We knew.

DAVE BALL, *Air personality*

I had doomsday clouds over my head from day one. When I signed the contract actually to do it, I was super-pumped, obviously. It was my first stint on air in a very steady manner, so I was ecstatic. The concept of it was so poorly executed, and so ambitious, that it didn't fit the traditional style; I didn't think that with the lack of radio veterans we had with us, that it was going to be something that was going to be able to be pulled off.

BRETT "SPIKE" ESKIN, *Assistant Program Director*

When I started there, I knew that Mancow was going away and I was very nervous because I didn't know what the plan was—even a month into being there, I had no idea what the plan was. Mike Stern sat me down in the office and showed me his PowerPoint that he showed Emmis and told me what the plan was. And the plan made a lot of sense to me. You say *Daily Show* ... that show came from the right place. Every element of it made sense ... as an idea. So heading into the idea, at first, I was fully in support of it. When they were doing the practice shows—so, (do) you ever, like ... you know how you have your favorite bands, right? And your favorite band comes out with a new album and you take home that new album and you listen to it? And every part of you knows that the album isn't any good, but you like the band so much and you've been listening to the band for so long and you have so much invested emotionally and mentally in the band that you keep listening to it to find the things that you enjoy? And you eventually convince yourself that the album is good, even if it isn't? —So, we were doing the practice shows and I was hearing the practice shows, and I knew I wouldn't listen to this show. But I liked the concept so much and Mike sold me on the concept so much, and I liked the idea of it and I liked that we were reinventing what morning radio should be; like, this is what morning radio should be. We should invest in ideas. I kept listening, until I found the little parts that made me laugh or that I enjoyed. Even as the show went on, it was the same thing. Whether it was Backseat Goat or "Hey There, Cicada," whatever little thing they did that worked, that caught fire, I would fool myself into thinking that it meant that it was viable. But really, deep down, I knew that nobody liked it, that they did things that people thought were not funny. Every part in that show, every person involved in that show, was a funny, creative, ridiculously talented person put into a situation and a framework that just didn't work. It was wrong. There was too much pressure on it to succeed too quickly. If it was on a cable channel, like an extra channel—it was too big a priority to be as unsuccessful as it was. I knew it wasn't good, but what are you going to do? This was it. We hired everybody. This was the idea. If it wasn't this idea, then what?

DAVE BALL

Hearing those practice rounds ... I was just like, "We are not ready to go live; I'll tell you that much."

NED SPINDLE, *Air personality*

I did have some worries about the non-radio people coming in from a completely different world, but I also had worries about the radio people trying to make it too much a standard radio show, when I really thought it could have been something that wasn't typical radio. It wasn't as unusual or groundbreaking as I'd hoped. I knew fairly early that it wasn't going to last, much earlier than the non-radio people did. Having worked at many radio stations, and seen many of them blow up, I saw the writing on the wall.

ALAN COX

We had Stephen Colbert on the premiere day of *the Morning Fix*. We were trying to get people interested in any way we could. Mike McCarthy, who had gone through Second City and SNL, had a lot of connections like that, so we had Colbert call in the show. That first day, we were trying to explain the crux of the show, simultaneously, to him and to the audience. I think it seemed strangely conceived in a lot of people's minds, so we said, "With all due respect, we're doing a *Daily Show* kind of radio thing." He asked, "How long is the show?"

I said, "We have four hours a day," and he said, "Oh, you're so screwed. I have to do 22 minutes." I said, "Oh, I know," and kind of jousted with him a little bit, and kind of went into my thing about how radio is the lowest rung on the entertainment ladder, but done well, requires the most work.

He was just, like, "I can't believe you guys have to do this every day for four hours."

He was incredulous as to what was going on. It was interesting to me, because obviously it proved prophetic, unbeknownst to us. I knew in my heart of hearts that the show we were trying to do was not going to be the blockbuster they hoped that it would be. When he said, "Oh you're so screwed," I guess we all collectively laughed, all 45 of us that were on that show. That was the very first day of the show.

STEVE TINGLE

I envy (Alan's) vocabulary. I remember when he first started working there, we were walking along Wrigleyville, and he went, "This is such a bucolic neighborhood." I looked at him and said, "What? What does that mean?" He went, "Bucolic, like beautiful."

ALAN COX

You've got to own what you are. I always have people giving me a hard time for my vocabulary, or whatever. I talk the way I talk. I'm kind of like, "God forbid professional communicators can speak."

CHRISTINE "ELECTRA" PAWLAK, *Air personality*

I wanted *the Morning Fix* to work for a couple of reasons. I thought it was smart, I thought it had a lot of potential, and I knew that Mike Stern believed in it. I didn't think Mike Stern would have taken the kind of risks that he did to put *the Morning Fix* on the air if he didn't really think that it could work. There were a lot of people involved in *the Morning Fix* who had no idea how to do radio. Doing radio is not as easy as reading something you've written. You can be a great writer and be a great radio host. You can be those two things independently. I think *the Morning Fix* didn't have enough people who could do both. I think it suffered from a lack of focus and a lack of guidance. Mike Stern couldn't be with them 24/7 to explain to them what his vision was.

ALAN COX

Initial public reaction was "Nice try, but this is never going to work." Robert Feder (media columnist) called it "bold but flawed," which I think still to this day is my favorite description of it.

JIM "JESUS" LYNAM

I was really rough on (the non-radio cast members) in the beginning. I would be, like, "This sucks. No, I won't do this," because I didn't want to have my name on certain things. I'm like, "This is going to kill me." As we went along, I realized that Michael McCarthy was a funny guy, James Engel was a funny guy, Ginger Jordan did a good job with news,—they were just given an impossible task.

KEVIN MANNO, *Air personality*

I was optimistic, the way that everybody talked about it. I was really optimistic. I was, like, "Wow, this has never been done before." I heard they were getting Michael McCarthy from SNL, and I thought it was really going to be professional, but it just was not. After the show, we would have a meeting every day to plan the next big show, and it would just be like four hours of screaming. I quickly realized that it just wasn't going to work.

ALAN COX

They always say that the best ensemble shows are built out of people who each know their roles, and know them really well, and (that) those (roles) are clearly defined. I think, if there was an Achilles heel of *the Morning Fix*, it was that it was kind of a death by 1000 cuts. We had to strike a weird balance between democracy and hierarchy. Michael McCarthy had been hired to be our head writer, but I was hired to be the host, to make sure that the whole thing ran properly, and to make sure that everything was done properly. Michael's a very smart guy, a very talented guy, but he had no experience in radio; there's just a different culture there. Our idea was to make it smart, but there were times where we were concerned about making it too smart. It sounds insulting to the audience, but everybody in radio knows exactly what that means. There were a lot of arguments among Dave Ball, Ginger Jordan, James Engel, and Jim, who was the lone holdover from Mancow. We would pitch ideas every day—and you're talking about a dozen people. I think Jim and I kind of—it ended up being just us. I think Jim and I were kind of viewed as being on our island by some of the other people, and I really tried to keep a balance. It just kind of ended up being a little trouble.

JIM "JESUS" LYNAM
Lots of people. Lots of people. Too many people.

SHERMAN, *Air personality*
It was not Q101. It was a horrible thing for them to try and replace Mancow. I think they had a lot going against them. I would not have wanted to take that spot. It would have been tough even for us to take that spot. Whoever replaced Mancow was going to have a really big uphill climb, and that show did not have all-terrain wheels to get up that hill. That show did not fit Q101. It didn't fit that lifestyle. It might've fit XRT, but at that time, it did not fit Q101.

NED SPINDLE
It was the funnest two years of radio I've ever done. I didn't have to sell anything. I didn't have to promote anything. I didn't have to mention any sponsors. All I had to do was write, produce, voice, and even sing comedy. Some of it was good; some of it wasn't so good. It was like a mini-Saturday Night Live. You'd have a meeting where you'd come up with ideas, and everyone would go off and write. It was fun, and I learned a lot.

JIM "JESUS" LYNAM
They brought us in and showed us this graph. It was, basically, like a big cross; it was groups of people: "There's 18-34 males, women, blah, blah, blah." So, then

they circle in this corner; it's like a corner, where we want all the people who are listening in, like, Starbucks, and listening to public radio. And I said it out loud. I said, "You've gotta be fucking kidding me." I got yelled at later on, but I couldn't believe—I went, "I've been on the craziest morning show maybe in Chicago history, and now they want to pretend like that completely didn't exist, and we're going to go after people who sit in Starbucks and listen to public radio." I thought it was insane. I said, "How would we do that? We don't even play that music." I was, like, "This is the craziest thing I ever heard." And that's when I started drinking.

STEVE TINGLE

I admired the concept—for Mike Stern actually to do it and try something different—that was ballsy. And if it worked? Hallelujah! I was part of something that was groundbreaking, but my gut instinct when I did it was, like, "Man I don't know how this is going to translate to radio, and I don't know if people are going to buy it or not."

JIM "JESUS" LYNAM

It wasn't like I was some great shakes, but I had an opinion and I made people feel something. Otherwise, you just had these people who were telling jokes and writing bits, and they weren't knock-them-out-of-the-park funny. And if you're going to do a show like that, every bit, every news report, everything you do has got to be an A+, because it's four hours. Their idea was people only tuned in the radio for half-hour spurts, so they were never going to hear it repeat. What they found out was (that) traffic in Chicago is *fucked*. People are sitting in their cars for an hour and forty-five minutes, two hours, two and a half hours—do you know what they hear? The same thing repeated. To me, that's where it broke down.

MIKE STERN

We knew going in that we weren't going to be able to do four hours of that type of content. The goal was 90 minutes, with some of that being interviews and music. When you took out music and commercials, it was significantly less than 90 minutes. The goal was to have three half hour blocks of content, and then they would repeat. The reality of radio is that people don't listen for four straight hours to a morning show.

JIM "JESUS" LYNAM

One of my sticking points was (that) people have got to be able to touch you and feel you—not literally, but they need to be able to call you and talk to you.

We completely removed that from a show. Without people involved, you, basically, become news radio. You're news radio, you don't take calls, you don't do nothing. But we were going to be really, really, funny. Even Jon Stewart is not funny, couldn't be funny, for four hours. How the eff were we going pull off being funny for four hours, without being able to talk to people?

MIKE STERN

(*The Morning Fix*) was frustrating as hell at times, and difficult and everything else, but how many program directors get to take a swipe at developing something completely new and different for the radio? Succeed or fail, I got that opportunity and I loved it.

MARV NYREN

I loved it. To this day, I still think it's a great morning show. In hindsight, I loved the concept of the morning show, kind of that Daily-esque Jon Stewart type of show. I just think it was wrong for that station at the time. There was an old line when Howard Stern was taken off a couple of stations; I heard so many people say, "You don't want to be the guy who replaces Howard Stern; you want to be the guy that replaces the replacement of Howard Stern." Whether it was (James VanOsdol) or whoever, I wish we had gone a year or two of being a music-focused radio station and seen how that may have worked. That may have been the right thing to do at the right time.

STEVE TINGLE

After a while, I knew that the success of *the Morning Fix* was going down, and I knew something was eventually going to change …

KEVIN MANNO

It's still a good concept, and I like all those people individually, but to put them all together, was just a bad show.

JIM "JESUS" LYNAM

One day, it was the end of the show and they let everybody walk out, and I think it was Tisa (Lasorte) who asked me to stay and said, "Hey, the *Morning Fix* from now on is you and Alan."

DAVE BALL

I walk into the studio office, and I see everybody with these sullen faces on. I'm, like, "What? We get fired?" And Ginger Jordan looks me in the face and says, "Yeah, we did." And I was, like, "Oh my God, no shit?" I just kind of laughed. I'm, like, "Well it's not like we didn't see it coming guys. Don't act like we all died. This isn't new news."

STEVE TINGLE

They wanted Alan and Jim in the morning, and they came to me. I think it was Marv. We were walking into the conference room; he went, "We think you and Sherman are going to work good together. We're going to team you up and put you in the afternoon." I didn't give a shit who I was going to work with, I was just so dying to be able to use my real voice and be me. And the characters could come into the show, but I didn't even care.

JIM "JESUS" LYMAN

It was what was. At the time, it was, like, well they just fired six people. I was lucky that I was there.

DAVE BALL

Management handled the exit like, "It's not because you guys suck; it's just a decision we have to make." Even though the reality was that, we sucked. I mean, come on, let's call a spade a spade here.

BRETT "SPIKE" ESKIN

To this day, I could tell you that what Jim didn't understand about me was that I understood Jim … was that Jim represented everything about the atmosphere and the workplace I was coming from—very bro, very masculine, very sexist, very homophobic. Even though these are pretty bad words, I don't mean them in a bad way. I was coming from a Howard Stern, Opie & Anthony, rock station in Philadelphia, a blue-collar town. Jim Lynam was more similar to me, and I understood (him) better, I felt like, than anybody else there. But I don't think Jim really got that I got him.

I think the clash I had with Jim was that he was led to believe, either by himself or by other people, that he was more important than I thought he was to the radio station. Jim's ego got really big. I was probably too young and inexperienced to know how to handle a guy like that the right way. Instead of treating him much the way that I would treat a program director, that I would trick him into doing what I wanted by making him think that it was his idea, instead of that, I just clashed with him.

Jim turned everything, much like a prototypic male does—turned everything into a contest, and it was him vs. everybody, and if he wasn't winning, then he was losing. And then, when the *Morning Fix* went away and just became him and Alan, he thought he was an even bigger star than he was, and he just became really difficult to deal with. I like Jim, at least on some weird level. I like Jim.

JIM "JESUS" LYNAM

When you're in that environment, I think I had a tendency to be a dick sometimes; I really did. You're fighting all day long, and you have a tendency to have this attitude. No one will know that I was lead clerk at Border's on Michigan Avenue. I'll always be the meathead jock dude. No one will know that I'm well read or anything like that. I was never going to be the lover of music; I was never going to be that dude. I had to do what I had to do; I had to play my role up to the hilt. Sometimes when you play that role 24/7, that role bleeds over into other things. Sometimes I wasn't the nicest guy in the world.

ALAN COX

The culture of the station was … I think it was safe to say that I had kind of missed the heyday of Q101. Q101 was no different from any other radio station, in the sense that the industry was under constant turmoil. I think if they had wanted to try that show even six months later, it never would've happened. I think I got into Q101 by the skin of my ass.

KEVIN MANNO

(Alan) and Jim Lynam had that show together, and I think that was sort of confused. I wish Alan had had a better opportunity at the radio station, because I liked him a lot.

TISA LASORTE, *Brand Manager*

A very unfortunate situation. Individually, talented men. Together, they did not make a strong show.

JIM "JESUS" LYNAM

It was a week before they started PPM, the meters. We get off the air; it was the day of Lollapalooza. They call us up there, and I see Alan come out of Tisa's office. I walk in, and I go, "What's going on?" They go, "Well, we're not going to renew Alan's contract. We're going to move Sherman and Tingle to mornings." I said, "Well, ratings come out in a week; why wouldn't you just wait to see what happens?" "Well, we've made the decision, we're moving on ...." I said, "Okay, well I guess I'll talk to you later." She goes, "Well where are you going?" I go, "We're done, right?" She goes, "Well, you've got to go work Lollapalooza." Swear to God, because I was under contract for another eight months. So I had to go down to Lollapalooza by myself and answer listener questions about where Alan was.

ALAN COX

The thing that bummed me out the most, and I don't mean to sound like a jag saying it, but that show would have been very successful if they had let Jim and I be ourselves. They were so focused on ... they were very much in the mindset of "we're going to win with music." This was the advent of PPM, so everybody was doing exactly what they said they wouldn't do, which was (to have) kneejerk reactions to PPM results.

When Jim and I left, we were #1 (with) men 18-34; we had a five share in PPM. So, they let us go with really good ratings, because the management at the time was so hell-bent on changing things up, and I think promises had been made to other people. Even within the massive confines of those two-minute segments, we did a good show.

THE LOST ART OF KEEPING A SECRET: LOOSE LIPS AND SUNKEN SHIPS

CHAPTER TWENTY-NINE

I was fired by Q101 on April 20, 2007, one day after Fook was let go, and only a couple of months after Program Director Mike Stern was fired.

Let me backtrack a bit. I returned to Q101 to host the soul-crushing and mind-fucking overnight slot in January 2006. When I came back to Q101, everything Thomas Wolfe warned about in You Can't Go Home Again *was made vividly clear—nothing at Q101 was as I remembered it. To generalize what I experienced, the station had become a brick-and-mortar Facebook. Gossip and petty insecurities were everywhere, and an alarming number of staffers seemed more interested in self-promotion than in actually advancing (or contributing to) the product.*

There was a choking sense of smuggery swaggering through the halls of Q101. Attitude was everywhere. Whenever I pass someone in the hallway—no matter where I'm working—I always make a point of saying "Hello" as I walk by. Common courtesy, right? When I would pass the much-younger members of the Marketing and Promotions team in the Q101 corridors, I'd rarely get a return smile, grunt, or wave when I passed. Nothing.

Because of the childish behavior around me, it shouldn't have surprised me as much as it did when I discovered that everyone at the station—everyone—knew that Mike Stern was going to be fired, weeks before he got the axe.

That scenario, disgusting as it was, seemed totally buttoned-up compared to my situation. The day I got fired, I found out it was going to happen hours beforehand from a listener who anonymously emailed me at 1 o'clock in the morning. In addition to blaming Stern's replacement, Spike, for the leak, my Deep Throat said, "This whole thing is bullshit! Sorry I had to break the news!" Yep, total bullshit. When Spike showed up at the studio door at 5:30 a.m., the first thing I said was, "This is the most fucking bush-league thing I've ever experienced." He said, "Yeah, I know. I'm not happy about how this all went down. Come on. Let's go to Marv's office." Fifteen minutes later, I did the walk of shame back to my car, aghast that I was on the street and that those in charge were still inside.

Can you imagine finding out that you were going to be fired via a friend of someone who worked in your office or place of employment? It was the most shaming moment of my career.

Back during my first go-round at Q101, if anyone had allowed sensitive information like that to leak to the staff ... and then to a listener ... they would have been shit-canned on the spot, no further discussion necessary. I know for a fact that Dave Richards would have suffered no fools in that sort of scenario. In my fantasy world, I imagine the no-nonsense PD saying in measured tones, "I want to know how this happened. I want to know who did this. And I want to know before the day is over. No excuses."

I'll never know for sure how the information about my firing leaked, and at this point, it doesn't really matter, but that didn't stop me from asking Spike when I interviewed him for this book.

Spike said, "I still, to this day, would love to know who the fuck ... like where the fuck that comes from. I think it all comes from salespeople. I think it must come from somewhere near the top, because I certainly wasn't talking to anybody about it. Every leak around always happens from the top to Sales. Somebody from the top tells somebody in Sales, and nobody in Sales can ever keep their mouth shut. Without naming names, there was one person at the top. There were two people above me who always knew when something was going to happen, and those people were people that were friendly with salespeople, and not me. The culture was very 'high-schooley', in that way. It was very 'everybody's sleeping with everybody, and everybody's gossiping about everybody.' I don't know if normal offices are like that, because I've never worked in a normal office. Both radio stations I've worked at have been like that. Nobody can ever keep his or her mouth shut about anything. I had no idea. I never talk to anybody but my boss about those things. The only thing that ever made me scared of getting fired was ... telling somebody something that I wasn't supposed to tell. That's what I was always scared about, so I never talked about those things. I have no idea. It must have come from the top, and it must have gone through Sales."

Several months after I was fired, I was asked to lunch by newly crowned Brand Manager, Tisa LaSorte. She said that the station had made a mistake in firing me, and asked if I'd consider coming back.

"No thank you," I told her. "That was the most disgusting, unprofessional, experience of my career, and I wouldn't go back to that environment for anything." She understood, and the meal remained cordial. But honestly …

I returned to the station to do some weekend shifts in 2009, but only after much of the station's infrastructure had turned over.

BRETT "SPIKE" ESKIN, *Program Director*

Everybody knew Mike Stern was getting fired except for me. I didn't know Mike Stern was getting fired. There was this dinner; we had dinner with Chevelle … Apparently everybody, including the record people at that table, knew that Mike Stern was getting fired, except for me. I had no idea. So, when Mike got fired, I thought I was getting fired too. I didn't know what was going on. I'd never had my boss get fired before.

Mike got let go, and there was just me for a while. And then they brought Tisa in as Brand Manager.

FOOK, *Air Personality*

I think, in the back of my head, (my firing) wasn't totally unexpected. Mike Stern had gotten fired three months before that. I had lunch with him probably about a month prior to that, and said, "I think I'm probably going to be next." I meant it in a half-joking fashion, but there's…no matter how many times you've been fired, there's no substitute for being freshly fired.

Spike did the firing. If I had been Spike, I would have done it very differently. As soon as I was done that day, as soon as I got the call—it was a call on the hotline—you know. "Hey man, can you come in here? I've got something to discuss with you." I knew I was getting canned. I walked in there, and there is … it was (Marv Nyren's assistant), actually. She was in there because they needed to have a witness to have you fired at the time. Spike's like, "Look, man, we're not going to renew your contract. I need you to get your stuff out of your locker and go."

I think I said, "Really? *Really*? Is this it? Is this how it's going to fucking go down? Just like this?" He was, like, "Yeah, this is pretty much how it's going to go down." I went, "Fucking God … great. This is fucking great." I stood up, twirled around with my arms in the air, and said, "This is fucking great."

BRETT "SPIKE" ESKIN

In my recollection, it's close, but there was less rage and there was no (assistant). What happened with Fook ... Fook came; his show ended at 7:00 but he probably came in at, like, 10 of, so she wasn't in there yet. He comes in, he looks at me, and he goes, something to the effect of, "Aw, am I getting fired?" And I was, like, "We're not renewing your contract." He goes, "Really?" And I said, "Yeah, dude, I'm sorry." Then he just took a deep breath and he was like, something to the effect of, "Well, that's fucking great." There wasn't arm waving. I didn't tell him to get his stuff. I think he asked if he needed to get his stuff, and I said, "No, man." I was, like, "I'm around. Just call me this weekend if you want to come in." And he said, "Okay," and he left. He texted me or called me that Saturday. I met him at the Merchandise Mart; he got his stuff, and then he left.

FOOK

Basically, three weeks before my contract was officially up, they told me to leave the building.

BRETT "SPIKE" ESKIN

The only reason I remember it that way is because it was just me in there, and I remember (Fook) being the easiest let-go that I ever had. James was certainly not easy. The *Morning Fix* thing was a disaster, even though they knew it was coming to some extent. Fook just kind of knew I felt like; like, he said he did.

FOOK

I certainly don't blame Q101 for firing me. I certainly wasn't doing anything to save that radio station, and a lot better talent than me came on after that and couldn't save the radio station.

MARV NYREN, *General Manager*

Working for a company like Emmis has some wonderful benefits to it. One of them is if you write a plan that looks good, sounds good, feels good, (and) tastes good, nine times out of ten, they let you do it. There aren't a lot of companies that let you do that. It also means, when you think about it, you're going to make mistakes. Morning Fix in hindsight, loved it, but it was a mistake for the time on the station. I also really changed some things up, where I kind of eliminated the traditional Program Director roles. We had two guys, Mike Stern and Tim Dukes

(WLUP Program Director) —two wonderfully talented guys, unbelievably nice guys; I can't say enough nice things about them as people—but at that point in time, I felt like we needed to do something differently. The market revenue was sliding; we had changed Mancow just a little while ago; the alternative product was at an all-time low; it was really struggling. So I changed up, eliminated the traditional Program Director's role, and brought in a Brand Manager, a lady named Tisa LaSorte.

TISA LASORTE, *Brand Manager*

(The Brand Manager job title) was a gimmick, created by our GM. He thought it would make it sound like we were doing something innovative in a press release. Silly PR stunt.

THE BETA BRAND: Q101.1

CHAPTER THIRTY

Q101 had been known as "Q101" since the 1980s. As radio brands went in Chicago, Q101 was a familiar name with a lot of history.

In the late 00s, the station decided to change the station's name, essentially going all "New Coke" on it.

BRETT "SPIKE" ESKIN, *Program Director*

Essentially, I made all the music decisions; most of the talent decisions, (Tisa) was involved in, but she was there for brand management. Branding was the "new thing" in radio. One of the things I guess we had learned was that people, whatever people are looking at, that's what you need to brand as. That was a group discussion—that was me, that was Tisa, that was (consultant Dave) Beasing. We were all, like, "This would be a way to do something, to make an impression." So, that's when Q101.1 happened. So I will take partial, but not full, credit for that.

MARV NYREN, *General Manager*

One of (Tisa's) ideas was to rebrand these radio stations (Q101 and WLUP) moving forward. One of the things they brought up and proposed was "Let's hip it up, in a digital world, Q101."

Hence, the Q101.1, which is the actual dial position. At the time, it made sense. The internet exploded a few years before that. Everything was going digital; if you pushed a button in your car, it said "101.1;" if you logged on to our website and tried to listen to the stream, it was 101.1. I got it. It made sense at that point in time, but in hindsight, it was kind of like Coke changing to New Coke. We changed the formula. What I learned out of that, a brand is a brand … McDonald's golden arches should always be McDonald's golden arches. The Nike swoosh, I believe, will always be the Nike swoosh.

TISA LASORTE, *Brand Manager*

The digital/PPM age (led to the change to Q101.1). The brand had started to lose its luster, and what was important in the PPM world was for people to know your "address" for actual listening. There was confusion in the market with the successful 101.9 (WTMX).

BRETT "SPIKE" ESKIN

It was so many different things to so many different people. It kind of meant alternative music, but it didn't really mean that. It was so many different kinds of music. So, here was this famous, but mixed-up brand that was meaning less and less to people; so, I don't think it hurt anything, and I don't think it helped anything.

AJ COX RUDLOPH, *Sales Promotion Manager*

It threw everything off. It didn't make a lot of sense. We were moving into PPM right then, anyway, so the recall thing wasn't such a big deal. And Q101 was a brand. It was a brand that was built up over time and it was Q101. You just don't change that kind of thing.

TISA LASORTE

In hindsight, I still believe it was the right thing to do, even though a lot of people associated with the station saw it as "killing the brand." I believed the brand had already lost its strength. It reflected in the ratings. And we needed all the help we could get to gain ratings while not trying to live just off old glory. Many will disagree. I know.

MARV NYREN

To alter it permanently was a mistake, in hindsight. The reasoning we (used to do) it, to be more current and digital, made sense at the time. Q101 should have stayed Q101.

SHERMAN AND TINGLE

CHAPTER THIRTY-ONE

Brian Sherman signed on with Q101 in early 2001, and paid his figurative dues over the next several years. His on-air status changed, and his brand skyrocketed, when he was paired up with Morning Fix *cast-off Steve Tingle, to host their two-man comedy/music show.*

SHERMAN, *Air Personality*

I was doing nights and the Program Director said, "Get a tape ready, "and whenever Program Directors say, "Get a tape ready," that means that they have some kind of idea where they want to place you. There's something up that they now need some audio of you to play for some other people to make this idea go through. So I'm, like, "Okay, something's up." So I give the PD the tape and after the tape, they come to me and go, "We have an idea for you—we want to put you on in afternoons with Steve Tingle." I didn't know Steve at all; I knew that he was on the *Morning Fix* and he did a character named Clarissa Jenkins. That's all that he did; he never did anything as himself on that show. He was just that character doing traffic. I didn't know him. I didn't know his beliefs on radio. I didn't know his personality. I didn't know how this would work. Ninety-nine per cent of the time, when they throw you together with somebody, it doesn't pan out. I was with a girl one time doing a morning show in Kenosha, and it was hell because we had nothing in common. We didn't get each other. There was no timing at all; it just didn't work out. But I was doing nights and I didn't really want to do that anymore; I kind of wanted to do a show. I didn't just want to be a jock introducing a song anymore; I wanted to do more. I knew that doing afternoons would allow me to do that. I said, "What the hell? Let's give it a try." So Tingle … that night, we talked on the phone for three hours, just about our ideas and what we believed, how he thought the show should go and how I thought the show should go.

STEVE TINGLE, *Air Personality*

We talked on the phone, and immediately, when you meet someone, you can sense their work ethic right off the bat, on what they do. I knew that he was the type of person, like me, where our show was from three to seven, but our show started at 11 and ended at nine, and that's how he was too.

SHERMAN

It worked out from the beginning; it's very rare when that happens.

STEVE TINGLE

When I was on the air with him, it was perfect. He was a great bus driver, and I thought I was a great passenger on his bus.

SHERMAN

It worked out great—he was one of those guys you run into at a wedding, you know what I mean? You go to a wedding sometimes with your wife—you don't know a single person, you didn't want to go to this wedding, (and) all of a sudden, you sit down at this table with some other couple across the way. You're drinking, (and) all of a sudden, it's like your long-lost brother's this guy you don't even freaking know at this table. You have a blast, and you get hammered all that night. That's the same way it was with him, minus the wedding.

STEVE TINGLE

When Alan and Jim were getting let go, they wanted to do a change in the morning. They came to us. They said, "We want to put you in morning drive." We didn't want to do it, simply because we were having so much fun and we loved the hours. We were living regular lives. We'd wake up whenever we wanted to … go to bed, not worrying about having to set the alarm for 2:30 in the morning. It was heaven. We didn't want to leave afternoons because it was heaven.

SHERMAN

So we're doing afternoons. Everything was going good; we were number one at the time; everything was great. We didn't have to wake up early; we didn't have to go to bed early. Perfect. Everything was great. All of a sudden, (the boss) comes up to us and goes, "Hey, we have an idea. We want to put you in mornings." This was (Program Director) Marc Young and Tisa, mainly Tisa. We go, "Okay, what are we going to do here?" So, we start weighing the options ... we're, like, "What do we do here? It's morning radio in Chicago." Then we're, like, "Yeah, but we're doing so well in afternoons. We're kind of like starting from zero again, going to mornings. We're totally changing our listeners."

And we were under the impression, too, that (there) wasn't going to be much more with it—like, the money aspect, either. At the end of the day, we said, "You know what? Let's just turn it down." So, we go into this meeting, and we sit down. There are two people in this meeting and they are convinced that we're just going to come in and say "Yes." So, we come in there, and we go, "You know what? We really appreciate this opportunity, but at this time, we're going to have to decline." Their faces dropped.

STEVE TINGLE

They said to us, if you don't take this, it's the biggest mistake of your lives. Marv came in and said, the exact words, "If you don't do this, it's your biggest mistake." I've worked with Marv for a long time, and I've never seen him mad like that.

SHERMAN

It was not like a power trip. We weren't trying to throw our weight around. We just, at that time, were, like, maybe we don't want to do this, maybe this isn't the right call for us at this time. We weren't trying to be dicks in any way; we just didn't think it was the right move for us at that time. So, we go back to our office. Everything's fine: "Okay, yeah, we respect your decision." We're in our office—I shit you not—five minutes. All of a sudden, the door comes swinging open. It's Marv Nyren, our GM, who goes, "You two guys made the biggest fucking mistake in your lives. You guys are never going to work in this town again. I hope you guys are fucking happy with what you guys have decided to do," and slams the door. Tingle and I look at each other, like,"Well, shit. I guess we're out of a job now. (We went from) turning down a promotion to fucking unemployed." So now, it's a mad dash to find all of our flash drives and hard drives, to get everything off the computers. We were convinced we were going to get fired in ten minutes. We're trying to get all of our emails, because we think the Microsoft Outlook is going to be shut off in five minutes.

MARV NYREN, *General Manager*

I honestly don't remember the words, but those might've been pretty close. What blew me away, what they may not have told you, (was) we talked a little bit about their future and where we could go, and I said, "Boy, I would love to see you guys continue to grow. I think mornings would be phenomenal because of all the elements that you have; I'm not sure we're going to be able to do those in afternoon drive. If the opportunity ever arose, we'd love to put you guys in mornings." A few months later, all of a sudden, we had this opportunity. I think they had been thinking about it, because they immediately said, "We're not sure we want to do that." I'm sitting there looking at them, waiting for a punch line … I've known Tingle for years now … I remember waiting for a punch line, literally, and none came. It was, like, "Okay guys, we'll talk later," and they got up and left my office. I sat there, and I got madder, and madder, and madder. Smoke started coming out of my ears.

I got up. I went into their office, I slammed the door, and I truly don't remember the words, but I said, "Guys, this is the biggest mistake of your careers: to turn down mornings in Chicago at Q101. What are you, idiots? There's more money involved in it, you're going to have a larger platform, you will be the faces and voices of this radio station." The other day-parts are unbelievably important, but every day starts with morning drive on every radio station. No matter how big the other day-parts are, if you don't have a good morning show, it's tough to be real successful. They were probably scared first, like, "Uh-oh, if we don't do this are we going to have a job?" And I also do think they sat down, and said, "Hey, you know what?" —maybe they didn't want to get up at three o'clock in the morning—but I think they realized, to do mornings in Chicago, for two guys in their early 30s, (who) didn't have that much experience …

STEVE TINGLE

It was almost like, he was mad; but I genuinely think he was looking out for our best interests at the same time.

MARV NYREN

They did it, and I think they had some phenomenal success—really well received by the advertising community. People loved them. They would work really hard. If there was an event we needed them to go to, they did it.

Sherman's good at it, but I think one of Tingle's ultimate destinations is doing something in front of a crowd. He is a performer.

ERIC "SHARK" OLSON, *Air Personality*

Had they "existed" 15 years ago, I think Sherman and Tingle were the right match for what we were looking for in the 90s. Had the station, back in the day, found a pair like them and stuck with them by letting them grow, and then, stayed away from the heavy rock it started to play, I think Q101 would still be around today. Sherman and Tingle became the Kevin and Bean (KROQ morning hosts) of Chicago, and it's very sad their run was cut short.

ICKY THUMPING: Q101 VS. THE WHITE STRIPES

CHAPTER THIRTY-TWO

The illegal downloading of music was a hot topic in the 00s, and constantly in the music industry's crosshairs. Radio towed the corporate line for years, staying above the conversation and practice, and sticking to outdated methods of music delivery for its on-air product.

Before the 2000s, music was sent or hand-delivered to radio stations by record promoters working for major (and, less frequently, independent) labels. Radio dutifully played along with label-recommended "release" and "impact" dates, withholding airplay and reporting like good little soldiers, in the hopes of courting and maintaining favor with the labels.

Spike, who was programming Q101 in 2007, took a chance and broke bad on the old radio and records model.

BRETT "SPIKE" ESKIN, *Program Director*
The Jack White thing was the most unreal thing that ever happened to me. I never had a rock star directly or indirectly mad at me before.

So we were in a stage where Mike Stern had been fired, and they did a restructuring. Basically, I was programming the radio station. Every bit of pent-up programmer angst I had from the past eight years—I got to do everything that I thought was the right thing to do—not all of these things were the right things to do—but when nobody's stopping you from doing these things, you do all of them.

One of the things that always bothered me was when labels told us not to play leaks. At this point, it was when everything was, like, "It's on YouTube. It's got, like, three million views, so people have heard it. Can I just not look foolish and play it?" We were just playing leaks. So when I got a record, we'd just play it. Nine Inch Nails: their record leaked. We just played it. Linkin Park leaked and we played the album like four times in one day. We made sure never to trash it, but I wanted to play it as quickly as the internet.

I was never good at stealing these things myself; I just knew enough people that scoured the internet for stuff; they would just send me emails. I got an email from my brother, who said, "Hey, this guy just sent me this link on this blog to download the new White Stripes record." I went and downloaded the White Stripes record—the blog wasn't even in English. I listened to it and was, like, "Yep, that's the White Stripes. Nope, never heard it before. Yep, must be the new record." So I called Electra and I was, like, "Hey, I've got the new White Stripes record. I want to play it." She's, like, "You sure it's right?" "Yep, I'm sure it's right."

CHRISTINE "ELECTRA" PAWLAK, *Air Personality*

We had quote unquote leaked a few records, most of the time with permission. (Spike) let the record label know that this was happening. For my part of it, I was excited, and giddy, and a fan of the music. I played it and talked after almost every song and before every song, talking about how I couldn't wait for the record to come out.

It was nothing but positive, from me. Then we got the call from Jack White.

BRETT "SPIKE" ESKIN

She played the whole (White Stripes) record, beginning to end. While this happened, I would call the record label, as we were about to do, just so they had the heads up. I'd be like, "Hey, man, I have this record. It leaked on the internet… just so you know this is going up."

He would go, "Well, I know I'm not going to be able to stop you."

Electra's in my office after her shift is over. The receptionist calls and says, "Hey, somebody's on the phone for Electra." I'm, like, "Who?" "Jack White's on the phone."

I laughed and said, "Jack White? What?" "He said he's calling from Spain, and he wants to talk to Electra."

CHRISTINE "ELECTRA" PAWLAK

I think Spike and I made a good team, in that Spike would frequently come in the studio and be, like, "Aw, you have to play this!" And I'd say, "Okay," or "Really?" or "Come on, man." And he'd say, "Aw, no, you've gotta do it."

And eventually he would win and I would play it. That was what we did; that was how we worked, because he was the boss, and I was a good soldier. So, we were in his office, and we heard this all-page that Jack White was on the phone, and wanted to talk to me. And I remember my heart just sank, and I thought, "Oh my God, really?" I walked into the studio where Sherman and Tingle were on the air at the time. And I was so terrified that I kind of let them field the call, but Jack wanted to talk to me.

BRETT "SPIKE" ESKIN

The first thing going through my head was that Jack White had heard about this I figured he thinks it's cool. In my head, Jack White's some anti-establishment, like, "It's all about the music," guy. I laugh, and I go, "Throw it to Sherman and Tingle, because they're on the air now." I tell Electra to go in there.

She gets in there, and I finish something up in my office. I walk into the studio five minutes later. Sherman, Tingle, and Electra's faces are white, like in shock. They were like, "That was really Jack White, and he was really, really, pissed off."

STEVE TINGLE, *Air Personality*

At first, I didn't believe it … and then when he came on, and he started barking, and Electra was in there, I was, like … I'm just not used to that kind of arrogance. It was very uncomfortable. And then he started bashing—I was getting fired up, getting ready for a fight. I started shaking because the adrenalin was coming in, and I just wanted to go crazy on the guy. I didn't know what my limits were. I was still kind of green behind the ears, and a little bit hesitant, but as soon as he started barking at Electra, I was like, "Eff You."

BRETT "SPIKE" ESKIN

They played the call back to me, and Jack White was furious; he was really angry that we had done that.

CHRISTINE "ELECTRA" PAWLAK

I remember that he closed the conversation by saying, "Merry Christmas; Happy New Year!" just dripping with sarcasm. And I remember that he wanted to talk to me and wanted me to take responsibility for what I had done. Other than that, it's kind of a blur.

BRETT "SPIKE" ESKIN

I talked to (White's) manager, who was really pissed off. The label wouldn't talk to me for the rest of the day. And then, Electra went home and wrote a blog about it. When we got to work the next day, MTV.com had it on their front page. NME had it on their front page. *The Reader* wanted to talk. MTV was calling me out by name. The BBC wanted to talk to us. Spin wanted to talk to us. It was the craziest thing I'd ever seen.

CHRISTINE "ELECTRA" PAWLAK

The gist of it was that he felt, like, not that we had stolen his art from him, but that we stole money from him. And I felt awful, really awful. But I'm a writer, and I went home and I blogged about it. And I didn't really think about what I was writing. I wrote very quickly on my blog, which at the time was called "The Queen of Snark" (I know). I outlined what had happened with a little bit of hyperbole and paraphrasing, but essentially relating it as I remembered it—that I played the record, that Jack White called me and yelled at me for playing the record, (that) it was surreal, and that I was still a huge fan of the music—that was why I played it. I have related this story to communications classes and radio students as a cautionary tale—I view it as a cautionary tale—what you write on the internet lives forever. The next day, one little music blog picked it up, then another, and then, by the weekend, it was on MTV's website, and *Spin*'s website. And I was something I had never wanted to be—labeled. I was the girl who Jack White yelled at. And I felt so helpless because I couldn't explain to all of these people in all of these states, in all of these countries, all over the world who thought that I had corrupted Jack White's art that, no, no, no, no, no, that's not what I was trying to do—I was trying to share it. That was all I wanted to do, was share it. Music is meant to be shared. That's why I wanted to be a DJ.

STEVE TINGLE

These people, their whole lives, they make music; and the reason they got into it is because one day they wanted to hear their song, and they wanted everyone to like their song. And for him to get mad for a radio station to play their song … I was like, "Screw you. You have totally looked beyond what the purpose of you getting into music is. You're in this business to make music for your happiness? Everyone's in it to make money. They're all in it to make money, or else they wouldn't do it."

BRETT "SPIKE" ESKIN

It was really difficult for Electra, but I thought the whole thing was kind of neat.

One thing that was funny. He never called the request line, and at the end of the phone call, he said that we couldn't air it. So we did a reenactment word-for-word of what he said.

CHRISTINE "ELECTRA" PAWLAK

(I got) horrible comments and emails from journalists and music fans, just condemning me for what I had done, a lot of times without an understanding that Spike had given me the CD, that the record label had been told of what was happening, that there was a whole behind-the-scenes element to it.

PORTRAIT OF AN AMERICAN FAMILY: THE MANNO BROTHERS

CHAPTER THIRTY-THREE

Ryan Manno successfully blazed a trail through Q101, as both the host of his own popular shows and as a cast member on Mancow's Morning Madhouse.

When it came to music, he was the voice of the street, being roughly the same age as the people who were listening to him, and showing up at a handful of live music events every week.

As part of Mancow's show, Ryan had to tweak his Q101 on-air persona a bit to fit in with the cast. And by "tweak," I mean, "become gay and Mexican."

RYAN MANNO, *Air personality*

When Freak left (for the Zone), Mancow said, "I want you on full-time."

I remember the exact conversation when he wanted me to take over Freak's role. He said, "I want you in, but instead of being the guy who was doing overnights and this role you were in, I want to change it up and I want you to be either gay or Mexican, what would you rather do?" There were a few days where I said "No." I said, "I do not want to do that." There were a few days, probably a week, where I wasn't on the show. He said, "All right, never mind."

I saw him again, and I think clearer heads prevailed. I saw him in the hall. We sat down, and he said, "I still really want you to be part of the show." I said, "I do, too, and I gave it some thought, and I'll be both gay and Mexican." I thought, "You know what, if I'm going to do it, embrace it."

From the time when Ryan's younger brother, Kevin, started at Q101 as an intern, many behind the scenes assumed that he'd be following in Ryan's footsteps.

KEVIN MANNO, *Air personality*

That was the only thing I wanted to do with my life; I wanted to work at Q101. That was it. That was my game plan.

AJ RUDOLPH, *Sales Promotion Manager*

Kevin actually started out in the Promotions Department. He was one of my Promotions Assistants for quite some time.

KEVIN MANNO

Ryan did everything a couple years before I did. I remember going when Ryan applied for his internship; we went to the Merchandise Mart together, and I sat in the lobby with him as he filled out his paperwork. That's when the excitement started, being there.

RYAN MANNO

I never felt territorial about (Kevin being on Q101), because honestly, Kevin, I always trusted to be ten times funnier than I will ever be. My faith in Kevin to do the things I can't do is through the roof. I recognize my weakness, and know that my weaknesses are his strengths in every way, shape, and form.

KEVIN MANNO

I was nervous (when I started out on the air); that's when I worried that people would judge me based on Ryan, like nepotism, "Who's this guy?" I still remember my first time that I opened the microphone; it was after "The Pot," by Tool. I was just, like, shaking, so nervous. I had done radio in college, but this was completely different, because at the time I still saw Q101 as this massive behemoth. (I was) still more of a fan boy than part of the team.

JEN "JAMESON" LONGAWA, *Air personality*

Kevin's more of a genuine person, and I think a lot of it has to do with Ryan being groomed through the Mancow end of the spectrum, and Kevin sort of doing everything on his own. Both are very talented though, of course.

KEVIN MANNO

I learned from Ryan, professionally and personally, probably more what not to do than what to do; I've sort of learned from his mistakes. He'll tell you the same thing. I'm a little more mature than he is, because I've learned from his mistakes and avoided them.

I wouldn't be here without him. He paved the way and made things a little bit easier for me.

SHERMAN, *Air personality*

Kevin is extremely talented. I think he's perfect for what he's doing now at MTV (hosting the since-canceled show *The Seven*). I think he fits much better at MTV than he did at Q101.

Shortly after Mancow's contract wasn't renewed in 2006, Ryan Manno landed in the night slot. His brother Kevin followed him in the overnight position.

KEVIN MANNO

I was doing overnights, and Ryan was doing nights. And then they overlapped two hours of our shows, so I would come in at 10. We would work from 10 to 12 together, and I would stay until 3:00 (a.m.).

RYAN MANNO

(Working with Kevin was a) dream come true. It's one of those things … the moment where you understand you really are going to be doing this show with your brother on this station you grew up listening to. You really feel like one of the luckiest people in the world. You almost don't want to tell people because you're afraid that you're going to sound like a dick for having your dream come true, but it was in every sense of the word a dream come true.

KEVIN MANNO

I think it was extremely natural … I think we had been working towards that for a while. It was probably what we wanted for years. I don't imagine I will ever work with anyone that I have such good chemistry. It's natural chemistry. We've been developing this chemistry for the last 28 years.

RYAN MANNO

Kevin and I have always gotten along so well as brothers. Yes, we did have some pretty heated ugly moments, as you do anytime you're doing something creative with anybody. I think there where shows where I would leave during the show because we would get into an argument about how something should go. I would literally just leave the show and have Kevin finish. He would call me on my drive home and I wouldn't answer, mid-show, just total bullshit.

ALAN COX, *Air personality*

Ryan and Kevin, those are two guys where, to me, it was only a matter of time before they went on to big stuff.

TIM VIRGIN, *Air personality*

I think Ryan Manno had more drive than I've ever seen out of somebody—I mean, both the Mannos; they both impressed me tremendously. They did what they did on their own. They learned how to market themselves from one of the greatest marketing people ever, and that's Cow. Cow's a marketing genius, whether you want to believe it or not.

KEVIN MANNO

Working with Ryan, when we did the night show together, and we did *Crash Test Radio* from 11 to 12, we had one full hour of control; those were the days. That was it. That's what I remember most fondly. If I could go back and relive any of it, I would like to go back and do that. We got to be creative, and then we got to also express our musical interests. It was perfect.

RYAN MANNO

I can't stress enough how completely free we were. No one ever said, "Don't talk more than this time, or you're talking too much." It was like, "Just … go."

KEVIN MANNO

I'm not sure why, but they let us play anything we wanted, as long as it was new. No one ever told us to play anything. That impacted me the most, and that's what I'm most proud of.

RYAN MANNO

I think knowing how proud of us (our parents) were, that they could have something to point to their friends or their co-workers and say, "Yeah, those are my boys," that always made me feel great.

ST. LOUIS IN CHICAGO

CHAPTER THIRTY-FOUR

From 2000 to 2009, the Q101 Program Director position was cursed. Dave Richards, Tim Richards, Mike Stern, Spike, Tisa LaSorte, and Marc Young had all come and gone. What would become Emmis' final programming decision for Q101 found the company repurposing their St. Louis radio programmers for double-duty in Chicago.

Tommy Mattern took on the Program Director roles for both KPNT ("The Point") in St. Louis and Q101. Kyle Guderian was elevated from St. Louis' Director of Digital Solutions to Operations Manager for Emmis Chicago, which included Q101 and WLUP.

Mattern stayed in St. Louis, and flew to Chicago on a biweekly basis. Guderian packed his life up and moved to the Windy City.

SHERMAN, *Air personality*

(Tommy) was the PD in St. Louis and Chicago—you can only be loyal to one thing at a time. It's like, if you've got two wives—you're not going to be banging both at the same time.

KYLE GUDERIAN, *Operations Manager*

Our first priority was to drop the Q101.1 logo and revert back to the original Q101. This was announced in the very first staff meeting. I hated that .1 from the first day I saw it.

POGO

When Tommy and Kyle took over, they also embraced me, and they were really cool. They made me feel more welcome, where I felt just detached during those years when I was doing (television) full-time.

KYLE GUDERIAN

Things Tommy and I immediately did: eliminate the use of music beds, eliminate talking into and out of stopsets, eliminate mandatory live liners, and eliminate set talk break times.

Getting rid of music beds was an attempt to tone down the "Top 40" sound the station had developed over the years. Also, as most jocks know, music beds can be a crutch that allows you to feel safe and often times prompts them to "park" and talk longer than needed. The only music beds they were allowed to use—including the morning show—was the intro of the next song on the playlist.

Not talking into or out of stopsets just made us get back to the music faster. Jocks always seem to take extra time to talk when they are going into a stopset. This only adds to longer periods without music.

The live liners were something we eliminated because we felt the jocks were always trying to sell the listener something. We asked them to be compelling and entertaining instead. They were encouraged to talk about station related things like Jamboree and Twisted, but never obligated.

Eliminating set talk times allowed the jocks to pick and choose when they wanted to create content each hour. Sometimes they would talk five times per hour and other times it was three times. I feel that doing these four things helped bring a consistent sound to Q101 that was more listener-focused.

Another thing that the new programming team did was bring back the Q101 heritage event "Jamboree," which hadn't been staged since 2002.

The first new Jamboree in eight years was a sell-out at the First Midwest Bank Amphitheatre, without having the benefit a clear-cut headliner. Three Days Grace became the de facto headliner on a line-up that included Puddle of Mudd, Saliva, OK Go, Seether, and Papa Roach.

KYLE GUDERIAN

We kept the ticket prices competitive, and for whatever reason, the show sold a lot of tickets, and we had a lot of sponsors. And I think in Chicago for that month,

Q101 was the highest-billing station in the NTR–nontraditional revenue, because of that show.

To be there and see the crowd and to know going into it that it was going to be a successful event, minus any rain, it was awesome. We were happy, and of course, we were immediately obligated to do more.

SHERMAN

The most support we had was from Tommy Mattern, when he first came in. He was really behind the show, and he had some good ideas. He wanted us just to talk, which was a relief. Cause everybody else at that time was always telling us, "Hey, it's got to be under ten minutes—get your shit out and get out." And he's, like, "Be yourselves. Talk about your stuff," which was what we wanted to do in the first place.

ALEX QUIGLEY, *Air personality*

I had been in afternoons or nights for, like, three months. And, actually the day that they, that St. Louis, took over, they flip-flopped me and the Mannos. (They) put them in afternoons and me in nights, without telling me why. And I think because I asked them why, when I finally tried to pin them down—"Hey, did I do something wrong?" —they didn't tell me why, or what I should expect to be doing ... just, "Let's do this." These are people's lives, man. I had a little girl at this point. It wasn't ... I'm not stupid at the way radio works. And you can get fired for no reason. There's no ... someone can say, "I don't like the way you sound. Bye." It's the way it was done; it was weird.

KYLE GUDERIAN

We had a clear vision—we wanted to make it a rock station, an alternative rock station, and it was about playing the songs that we felt would make the most sense to the (audience) we had. It wasn't about appealing to our egos, or to people that lived just in the city. Our approach was, we know who our listeners are; they're white, males, living in suburbia, way out there. A lot of them aren't in the city and that whole city vibe didn't seem to be its strength. And by that, I mean Tool, Rage, and the harder-edge music that's probably shunned by the hipsters, made a lot more sense for the dudes out in the suburbs. And that's what we did.

JEN JAMESON, *Air personality*

I only met Tommy once, ever. I think I remember what he looks like, although Kyle, I was able to set stuff up with.

I worked at Q101 from 2009 to 2011. Though I never spoke with Tommy on the phone or in person. I never had a problem reaching Kyle.

KYLE GUDERIAN

Tim Virgin was a guy I knew from the old Q101 days, as a listener. Because I'm from Minnesota and I worked in Ft Wayne, Indiana, I did a lot of commuting, and I would drive through the Chicago area. Tim was always one of my all-time favorite jocks, growing up.

He was one of my favorite jocks of all time. I worked with him a little bit when I worked at WEQX in Vermont as the Program Director, and he was my Columbia record rep. We never got to know each other really well; so, on a personal level, I was excited even to get to meet him and hang out with him.

TIM VIRGIN, *Air personality*

I was so nervous to go back on the air because it had been ten years. You're, like, holy fuck, that's a whole generation—are people going to remember me? Is it going to matter that people remember me? Are people going to be, like, fuck you, who are you?" But it was Tommy, and it was Kyle. It was these guys that I knew before; so, it was all awesome.

CHRISTINE "ELECTRA" PAWLAK, *Air personality*

(Tim)'s a big softie who would tell me "I love you" every day, give me a huge hug, and tell me about his dog, or his bike, or these stories about growing up in Cleveland, or … Tim Virgin has a story about everybody. He strikes you as this hard-partying, rock star kind of guy, but he's a human being just like any of the rest of us are.

TIM VIRGIN

I made it back. That's so fucked up. That doesn't happen.

KEVIN MANNO, *Air personality*

(Tim)'s like a 22-year-old trapped in a 42-year-old's body. I'm not a praying man, but I pray for Tim Virgin. I hope that he finds his way. I love that dude.

SHERMAN

Party fucking animal. I've never seen any fucking guy drink and perform on the air like that. I saw him after he got shitfaced with U2. I'm, like, "I don't know how you're doing it." Bono came in one time with a bottle of Jamo, and they drank the whole bottle. I'm like, "I don't know how you're fucking doing it." He said, "I don't know either!"

POGO, *Air personality*

I think he's the ultimate dude. I don't think anybody listens to him or wants to just go out and drink with him. It's funny because that laugh, that hyena laugh, and that enthusiasm … no one in our station's ever putting something on. We're just us. That's just him. That's him.

He's kind of always on, isn't he? That's why he's probably always been an afternoon guy for his whole career; it's that energy. When (people) are staring off into space like zombies on their way home in traffic, he's perfect for that spot.

TIM VIRGIN

To be a part of what James VanOsdol and what Robert Chase and (what) all those people that I've worked with that are legends in the city of Chicago (were), that's a proud thing to be a part of that.

PARTY HARD: DRUNK SHOWS

CHAPTER THIRTY-FIVE

Radio's one of the few industries where bad behavior isn't so much rewarded as it is tolerated. Case in point: Sherman and Tingle's "Drunk Shows."

STEVE TINGLE, *Air personality*

Ah, the world-famous drunk show. We had two of them. The first time, I think we were out doing a gig from 8:00 to 10:00, and afterwards, we hung out with the listeners and just got wasted. Our vibe on the station at the time was "Dude, this is a party. This is a party. This is fun. It's kinda reckless. You can kind of do whatever you want. Let's go in the radio station, turn it on, and do a show!"

SHERMAN, *Air personality*

We got rocked at the bar. It was a pub crawl. We knew we were going to come back to the radio station—there was no way we were going to drive. We were doing afternoons at the time ... and we came back, and we were just going to screw around and play a phone tap on the radio, because nobody was on Q101 in the overnights. They had replaced the overnight person and just made it a computer, because it's cheaper to do it.

MARV NYREN, *General Manager*

It happened, I don't remember the exact time; 2:00 in the morning? The next morning, I came in and someone said, "Hey, did you hear Sherman and Tingle last night?" I go, "I know they did a performance somewhere, but no, I didn't go." They said, "No, no, on air." I'm, like, "What are you talking about?" They said,

"Oh, they came and took over the radio station at, like, two o'clock in the morning, and they were hammered."

SHERMAN

We started playing phone taps. Then, before (we knew) it, we had our headphones in there, and then the mics were turning on. Now we were having fun; now we were taking calls. We had no intention on staying as long as we did. We were just going to jump on and do a break. Well, the break turned into about three hours. In our credit, we did play all the commercials.

STEVE TINGLE

When we were done, we thought, "Uh-oh, that probably wasn't a good idea."

SHERMAN

Everybody was very well behaved. Nobody said anything bad. Nobody swore. The listeners behaved themselves. So we went to bed; everything was fine.

Jim (Lynam) and Alan (Cox) came in, like, an hour later and they realized something was up because the computer was off on time, or something. They figured it out. And they noticed that we were still there; we woke up at like nine in the morning. They started talking about it. Nobody probably would have caught us, had they not said anything. So they talked about it on the air, about us going on hammered that night. The suits didn't even find out, because they weren't listening to Jim and Alan. Nobody found out for three days until somebody told them, "Hey, did you hear Jim and Alan talk about Sherman and Tingle the other day?"

MARV NYREN

I went and listened, and in my opinion, they were intoxicated.

I think now, they'll probably admit it. At the point in time, they said they weren't, that they had just a couple of drinks but they were sober when they did it. I called B.S. on that one. The issue I had there (went) back to General Manager Rule #1: Don't lose the license. There are written rules, and operating agreements … you're not supposed to have alcohol in a room, much less perform while you're intoxicated. And they, I believed, had done both in the previous evening.

SHERMAN

They really were not that upset. Marc (Young) and Tisa (LaSorte) basically said, "We're not pissed that you did it; we actually thought it was very funny, very entertaining. But, we cannot condone that you put the license in jeopardy by being on the air drunk; there's no way that we cannot reprimand you for this.

So they suspended us for three days without pay. It was the best three days without pay ever, because that got so many people talking about us, it was ridiculous. It was like wildfire. It spread on Facebook; people were talking about it in the industry—it was nuts.

We had to sign a contract that we'd never do it again.

STEVE TINGLE

We signed paperwork that we weren't going to do it again, but when the new Program Directors came in, they said, "Dude, that was awesome, that's street cred—you should play it back."

SHERMAN

Our current bosses that we had right before we got fired (in July 2011) heard the (first) drunk show; they thought it was hilarious. We had the impression—they will not agree—that it was okay to do it again. That that's just something you have to do without asking. I'll reiterate for the book—it was our impression. We did not have that in writing. Anybody in radio that's reading this book, be sure you've got it in writing before you go and do it again.

KYLE GUDERIAN, *Operations Manager*

If you ask them, there was a misunderstanding that we had given them the sort of nod to do it again if it ever organically happened. But my position was that was never the case, and if anything, they could have at least called to get approval. We probably wouldn't have approved it even then. We don't want to put our license in jeopardy.

SHERMAN

That takes us to about two years later; we're doing mornings now.

STEVE TINGLE

We're, like, "You know what? People liked it." We thought management was cool with it. So we did it again.

That one was a little sloppier, and a little more drunk. They were not happy because Sherman, I believe, (disconnected the station) for, like, an hour or two.

KYLE GUDERIAN

There was just a random night where I came in the next day, and I went to the Media Logger to pull some airchecks from some jocks and I noticed a bunch of overnight audio popping up.

I'm, like, "Geez, why is there, like, 54 minutes of audio at, like, 1:30 in the morning?" So I click on it, and lo and behold, here's these drunk guys talking on the radio. I'm, like, "Wow, they actually talked for 45 minutes without playing a song. So immediately, I reached out to Marv and Tommy and said, "There's an issue here." And we addressed it.

SHERMAN

Had we had it in writing, we would not have been worried about our jobs for about a week, thinking we were going to be fired. We went and did it again with the impression that it was okay to do it again. It was that next day (when) we got a call saying, "We heard you went on the air again. This is a huge deal."

MARV NYREN

It was kind of a second offense. We had a new president of the radio division, very new, and he kind of wanted to make a statement. Only two times in my career, in 29 years in this business, have I ever truly had to say, I need to call my reputation online with you, my boss, to ask you to do this for me. If it doesn't work, I understand the consequences. I said, "Pat (Walsh, COO), I believe in these guys enough—especially Tingle—that we need to bend the rule a little bit. I understand we need to make a statement. I fully get it. I think there's a way to make a statement without letting them go." I basically called in whatever chips I had accrued over my lifetime, and Pat said, "Okay, I just need to see what the disciplinary action plan will be, and I'll think about if I'll let you keep them or not." They were on probation forever, for the duration of their employment at Emmis. There were a couple of other things, I just can't remember what they were, that were written into their agreement. They had to sign it. They had to agree to it. I think they even took a minor pay cut. I think it was a pay cut, a non-paid suspension, (and) probation, literally for the rest of their lives, as long as they worked for Emmis. There were a couple other things that were written to that. They signed it, agreed to it, and never had another issue after that.

KYLE GUDERIAN

Jeff (Smulyan, CEO) was involved because Pat flew back to the States with Jeff on the plane from somewhere over in Europe. If you talk to Pat Walsh (about) his words in the meeting, which I sat in with Tommy, Pat Walsh, Marv, and the boys, Pat said to those guys, "Before I left, Jeff Smulyan looked at me and said, 'Just don't lose my … license.'" That was the marching order for Pat. It was a pretty serious meeting. I honestly don't think the guys understood how serious it was. They may now, reflecting back, but it could've ended a lot sooner for them.

SHERMAN

It went all the way to the top, to the owner of the radio station. Jeff Smulyan had to sign off. He was the decider; he was Simon Cowell on whether we were voted off. (He) said, "As long as they don't lose my license, they can stay."

STEVE TINGLE

The second one, we were this close to losing our jobs.

FOOK, *Air personality*

I learned more about Sherman after I got fired and he took over the morning show; I thought he did a really good job with Tingle.

MARV NYREN

As a corporate board person for the company who's in charge of the radio station, I certainly understood the severity of it. The saving grace of it was that it was two or three o'clock in the morning; we didn't have one complaint. I wasn't as concerned with what happened; my concern was really more about, "Guys, you can never, ever do this again. You really have done it twice. The first time, it kind of went unnoticed. The second time, it went noticed. We found out about it, and if it ever happens again, there won't be a conversation. It will be "Here's your stuff. Give me your key, and you're gone."

SHERMAN

We said, "Okay, that will absolutely never happen again. We will not speak of it. We couldn't talk about it on the air. We couldn't mention 'drunk show' after that; we couldn't say what happened. But now that everything's gone, it really doesn't matter.

INFINITE SADNESS: THE LAST MONTH

CHAPTER THIRTY-SIX

SHERMAN, *Air personality*

We had been hearing that the radio station was for sale for five, six, years. Oh yeah, might be sold … changing to country … changing to news. I had the luck of having my spouse also work for the company, so we'd kind of both hear things, and then, we'd just put our rumors together and kind of decipher which ones were true, and which ones were kind of stupid. When we heard of this last one going down, this was the one that I actually believed the most out of any of them. It's weird; I don't even know why. Actually, I told Tingle, too. I'm, like, "Hey, I heard a rumor that we're being sold, and they're going to announce on Friday." I heard this on a Monday, the week before it happened. But then, at the same time, I'm, like, "I heard this before. I'm not really that worried."

KYLE GUDERIAN, *Operations Manager*

I had suspicions that things were happening in November of 2010, just through some stuff I picked up amongst people talking. I remember it being around Thanksgiving, so it took a long time.

CHRISTINE "ELECTRA" PAWLAK, *Air personality*

I've never been on the business side of radio, but I knew (a sale) was possible.

TIM VIRGIN, *Air personality*

Nothing should surprise you in this business.

MARV NYREN, *General Manager*

Emmis owed a decent amount of money at high interest rates, and we were going to have a tough time paying it back. Therefore, Jeff (Smulyan) had made it clear that various stations and markets were going to be looked at. For various reasons, some were accurate in print; some were not.

KYLE GUDERIAN

For whatever reason, the actual sale happened really quickly, so I found out probably 24 hours before that meeting

At 4:06 p.m. on June 20, 2011, General Manager Marv Nyren fired off an email to the entire Q101 staff. The brief message asked that all personnel attend a mandatory meeting at 8 a.m. the following morning. The email closed with Nyren saying that he would be in meetings for the remainder of the day, and that he would be unavailable on phone and email.

The following morning, the assembled staff was told that Emmis Communications had sold Q101, along with sister station WLUP-FM (The Loop) in Chicago, and WRXP in New York City. The three stations were sold to the new broadcast company Merlin Media, who partnered with Chicago-based equity firm GTCR to acquire the group of stations for a reported $110 million to $130 million in cash, and $30 million to $50 million in preferred stock. The takeover, staffers learned, would happen no later than 45 days from the announcement.

MARV NYREN

When I sent the email, I didn't want to look people in the face and lie to them ... I said (to my assistant), here's the deal, I'm going to hit 'send.' You and I are going to walk out of the building." I didn't want to have to look somebody in the face and tell them a non-truth, or just not be able to talk to them. If there's bad news, I'd rather look somebody in the face and say, "Here's what's happening. It might not be what you want to hear, but here's what's happening." I hit send, ran out of the building, my phone blew up, and I truly didn't answer (it once).

TIM VIRGIN

He may as well have said, "Hey everybody, we're getting sold tomorrow." There was nothing more obvious. In my all years of being in radio, when somebody says, "It's a mandatory meeting"—I don't even (have to) know what it's about—it is usually something absolutely fucking horrible. Did I think it was going to be this fatal? I hoped that it was the Loop with all my heart, and I love the Loop. But I wanted our extension to be as long as the Loop's life was. I believe that people moved on from Led Zeppelin and now want the Pearl Jams and Rages and Smashing Pumpkins to be the Led Zeppelins, you know?

DANA LUCAS, *Air personality*
Within about a half hour of that email being sent out, we all knew the end was coming.

SHERMAN
As soon as that email came out, I'm like, "Okay, now it's happening." Any time they write that, and they're not answering the phone, that means they don't want to answer any questions that you have. I knew, especially at that moment, they were going to sell.

CHRISTINE "ELECTRA" PAWLAK, *Air personality*
I knew that was it. I got hammered that night. (I) went to a Sox game, ended up at some 3 a.m. bar playing pool, knowing that tomorrow everything would change. I didn't know whether or not I'd be losing my job … it was very difficult dragging myself in the next morning and thinking, Oh, perhaps I've made a grave, grave, error."

KYLE GUDERIAN
There was a whole lot of drinking the first week.

MARV NYREN
(The sale of) Q101 was somewhat of a surprise, but because we were in one location, one lease, we shared so many employees; trying to separate those to a buyer—someone has to buy either the Loop separately and Emmis keeps Q101—would have been nearly impossible. It would have cost so much money to split the stations. I think in the end, it was, literally, if we're going to sell one, we have to sell both.

CHRIS "PAYNE" MILLER, *Air personality*

I was surprised, but I knew the station had been for sale for a while. But I never dreamed anyone would buy it in this economy — I mean, what a shitty purchase. You know it's true.

SHERMAN

If this transaction happened a year ago, you'd be talking to (WLUP air personality) Pete McMurray right now, because the Loop would be gone, and that's the God's honest truth. Nothing against Pete McMurray — that's just the way the station was. It was a total reversal the last year, because the Loop wasn't doing well and we were doing great. Hell, we were number one, 18 to 34; we were doing great. The morning show was doing great; the station was doing great. And around fall, it kind of just petered … it all boiled down to, it petered out a little bit and the Loop got that momentum and had a little better ratings, and that's what happened.

TIM VIRGIN

Did I see this coming? I really thought Emmis, as a company, would've kept this one because this was their baby; you know what I'm saying? I understand how business works; at some point, you've got to let your kids go. Did I see it coming? Yes? Did I want to see it? I will deny that I wanted to see that station go away till the day I die.

There was a turnaround; we were on an upswing, and people were happy, and it was awesome. I just wanted Jeff Smulyan to walk through the hallways and go, "Wow! The vibe is great!" When the Loop was good, Q101 sucked balls. When Q101 was good, the Loop sucked balls. Nobody could ever get on the same page. That's when I knew, as a cluster, we were probably cluster-fucked, you know?

MARV NYREN

I think it had to be done to keep the company solvent and in a really good position possibly to grow in the next year or two.

SHERMAN

I really don't know who could've brought that thing around. I got the feeling it's been like that for a while. And it's sad to say, the station kind of was like the Titanic. It seemed like it had a hole in the boat for a while, and it was slowly picking up water. All the changes – it got so frustrating for us being on the air because we would get something going that was working really well, and then at that moment, somebody would have to change it, like, "Oh, you have to do this different now. Oh, I know that you're number one 18-34, but you've got to go after 18-49 numbers."

MARV NYREN

I think there was a lot of sad hearts that day. Also, during that announcement, it wasn't made clear that things were going to change, that Q101 was going to become FM News. We had heard a lot of rumblings; there had been conversations about it, but no one handed me a blueprint that said, "Here's what we're going to do to the radio stations."

CHRISTINE "ELECTRA" PAWLAK

In that first meeting, we heard that Emmis didn't have an idea about what the new owners were planning on doing. And within about an hour or two, everybody knew.

CHRIS "PAYNE" MILLER

It was the end days of Q101 and ... when I knew the end was there, I was just trying to stack interviews up. I was lining up the Smoking Popes—didn't happen. I was going to try to get Alkaline Trio on again; that didn't happen either. There were a few that we lined up that didn't happen, but one that did was getting Tim from Rise Against, and I really wanted that to happen. We were going to do a phoner—disappointing, but I'll take whatever I can get. About a week before it was scheduled, he was, like, "Hey, man I can come in if you want." I'm, like, "Really?" He's, like, "I'll be in town doing an interview for CNN on such and such date; can you do it then?" I'm, like, "Absofuckinglutely, are you kidding me?"

CHRISTINE "ELECTRA" PAWLAK

The hardest part of the last few weeks was the time between the end of the meeting in which we were told we were getting sold and the day three weeks later when I found out for sure that I was losing my job.

RYAN MANNO, *Air personality*

When the news came out that the station went away, people were reaching out almost like a relative died, like, "I'm sorry. What are your thoughts?"

CHRIS "PAYNE" MILLER

So, what I did immediately was the math. I'm, like, okay, 45 days. 45 days. All right, so that means the new station will take over August 1st. I flipped through the calendar really quick—August 1st is a Monday. That means the last show on Q101 is going to be *Local 101*. And then *Loveline*, but who gives a shit? So I call Jaime (Black); I'm, like, "Dude, there's a very strong possibility that *Local 101* will be the last (show on the station)." He said something the equivalent to "Holy shit," and I said, "Don't tell anybody. Let it just happen, because if they figure it out, they won't let it happen; they will not let me do it. Just don't tell anybody." So we started plotting and planning …

JAIME BLACK, *Producer (Local 101)*

I very clearly remember getting that call from Chris. And I absolutely shared his sentiment that, if *Local 101* was somehow going to be the last show on Q101, if that was how the schedule was going to play out, someone in charge would "fix" things to ensure it didn't happen. We knew the station flip was going to happen as a result of the sale, but we didn't know the exact date and time. So, my thinking was, even if Chris' prediction was right, it wouldn't be *Local 101* that shut down Q101. We're a one-hour specialty show buried on Sunday nights--we're not the song that closes out the final encore at the last show, so to speak.

Suffice to say, no one was more surprised about *Local 101* actually closing down the format than Chris and I were. And it's imperative to stress that it absolutely was an honor that neither Chris nor I took lightly.

CHRIS "PAYNE" MILLER

So, I guess that the (final) deal was done on a Friday. I guess that's when they had a five-day clause to take over, and then after five days, we'd, essentially, have to leave the premises. So, that apparently was exercised on Friday; so, then Saturday and Sunday of those five days happened, and then Monday, I was told by Jaime that the full-time staff was told that, "Thursday's your last night on the air; everyone's terminated as of now, but you're going to be on the air until Thursday."

CHRISTINE "ELECTRA" PAWLAK

On July 11, Kyle came into the studio and said, "Thursday's your last day." So what I did from that moment on was try my best to give Q101 and alternative radio a proper send-off in Chicago. I mean, I had to say goodbye myself and that was difficult because I had made a place for myself there for six years, but I wanted it to be that awesome, uniting, collective experience that new listeners and old listeners and people who'd drifted and people who'd always stayed could appreciate. And I came in with lists of songs every day. I burned CDs from my own collection to bring in. I got a lot more personal; I got a lot more emotional. I felt like all those years of being patient, of showing not telling, of building that foundation brick by brick came back to me tenfold that week. I realized that there were kids who graduated from high school, left Chicago to go to college, graduated, and came back and were happy that I was still on the air. People had met their fiancées and gotten married and had their first kids, and people had lived their lives with me being some small part of that. They appreciated that, and I appreciated them, and we weren't afraid to tell each other about it. It was three days of really honest radio for me. The sort of connection that I think has been lost.

The last days of Q101 were among the most memorable in the station's history. Once the official handoff date of July 15, 2011 was announced, the Q101 rulebook was torched. Operations Manager Kyle Guderian and Program Director Tommy Mattern allowed the talent to let loose on the air, giving them free rein to talk about the sale, and allowing them to play whatever songs they wanted to spin.

The full-time air talent: Sherman and Tingle, Electra, Tim Virgin, and Pogo, were open, honest, and engaging throughout Q101's final days. Without having to worry about formatting their talk breaks to conform to ratings-gaming best practices, and without a power-rotation playlist dumbing the music down, Q101 became unpredictable and exciting. It was a teachable moment for all commercial radio to learn from, one that sadly had to come from the demise of a long-running station in America's third largest radio market.

SHERMAN

I do give the company credit for allowing us that last week of radio. I was saying that on the last day … most times, they don't let you have that opportunity. When they sell the station, they say, "Give me your key fob, your computer is shut off—get out." I still give them credit for allowing us to do that, and for allowing us to be ourselves, to say what we wanted, to be real, to say just kind of what was going on , and what you had to look forward to.

POGO, *Air personality*

It's nice to be that honest with people, not that we weren't.

KYLE GUDERIAN

I knew that something special was happening. It was exciting, as a programmer and a listener to hear what was happening on the air. You immediately start thinking, "Why can't this sort of energy be pumping through the station at all times, and how do you tap into that when you move on to your next radio job?"

STEVE TINGLE

The final week was my favorite week on the radio there.

DANA LUCAS

I think I became a fan again in those final days.

KYLE GUDERIAN

I don't think we were ever huge sticklers on, like, must … stay … on …. playlist … at all times, unless it got really bizarre. At that point, all bets were off and we were just riding it out, and it was fun.

TIM VIRGIN

It sort of all just came together. Towards the end, we were allowed to play whatever we wanted.

SHERMAN

There were no strings on us at all; we could do whatever we wanted.

STEVE TINGLE

There were no rules; the program directors didn't care what music you played or how long you talked. You just got to be yourself, which is the greatest thing in radio, when you just get to be you.

POGO

There was no pimping some contest or ad copy, or something like that. I think I speak for everybody, we felt more for the listeners. Like, we'll all do something; we'll all find jobs. We'll all get around. But if you're that person that woke up every single day, tuned us in for years and years and years, and loved us, these are the people I feel bad for. It was nice to connect with these people.

MARV NYREN

What you heard was *real*. Tommy Mattern (and I) had a conversation about (whether) we muffle people? Do we not want them to talk about things? And both of us said, "Nah, this thing's going away. An era of Q101 is ending, away, basically, forever. Let's let them do what they want to do. Obviously, don't drink on air, don't swear on air—we never had any issues—well the last day, there might have been some drinking going on. I turned a blind eye, because I figured they kind of needed it. It was one of the most rewarding weeks, and I never would have thought it would have been, but it's when you see true passion for a product and for people. The final day I stood in that room on and off for twelve hours.

STEVE TINGLE

We were like, let's just have fun and go out with a bang and try to be happy. The only thing that got emotional was when we said our final goodbyes at the end. I was just really bummed out that I was not going to be working anymore with the guy who got me there, Marv Nyren—I've been working with that guy since I've been on radio. . And I was really bummed out that I'm not going to be working with that guy anymore, because he's such a fun person to work for; he doesn't micromanage. He's just a fun guy.

CHRIS "PAYNE" MILLER

I emailed Kyle … I'm like, do you think I just did my last *Local 101* show? He goes, "You probably did." And this is over an exchange of emails. And the Sunday before, I just didn't want that to be my last show. I just didn't want it to be the last show. It was a good one, whoever it was. They were all good in the end days, but I just didn't want that to be the last one, so I asked him, I said, "Do you think I can do one more show to sort of celebrate selfishly, not only my tenure at Q101, but my entire career as a radio host?"

I wanted to do it and just go out happy, and in style. He graciously came back and he said, "Absolutely. In fact, you know what? You've been on this station, been on the air longer than any of the other jocks. You host the local music show. It's only fitting that you do the last hour." So he brought it up. I didn't even have to ask for it. It was a dream come true. Honestly, a dream that came true.

TIM VIRGIN

The Q101 I know was from the time before I got there. Pre-Mancow to me is Q101, that's what it really is. Pre-Mancow and After-Mancow. The radio station was built on (those guys) who were really what alternative radio was. Everybody fucking talked about KROQ but to me it was always about Q101. I wanted to know how they were going to bury it. I wanted it to have its dignity, and nobody was doing anything about it, and it made me kind of fucking angry. I was, like, "Are you fucking kidding me? Like, you're just going to turn this fucker off?" At the same time, usually when a sale happens, they don't give you that chance. They usually lock the doors, say you're done, here's your check, and don't ever go on the air again. All of a sudden, the next day I went into work on the air, and I'm, like, "Holy fuck, I'm on the air; this is crazy. Do these guys care? If they don't care, let's fucking do something, you know? You can't just let this fucking thing go."

KYLE GUDERIAN

Tim Virgin, he really spearheaded everything (the last day).

TIM VIRGIN

I went into Kyle, like, "What can we do?" He's, like, "Well, what do you want to do?" I was, like, "I'd love to have everybody that's ever been on, on." He goes, "Yeah, that would be cool."

POGO

Tim and I immediately put the call out, like dispatch, boom; let's find everybody. And of course, they came out of the woodwork either to show up or to call in or to record something and send it in, and we decided, we've gotta make that last day about everybody.

TIM VIRGIN

How did I feel when I called and I invited Steve Fisher? I took the guy's job. I think that was probably the most emotional thing.

CHRIS "PAYNE" MILLER

I remember the night before, I was up all night long. I was up until about 3:30 in the morning, just writing a scratch list of bands that I wanted to try and play at the very end.

FOOK, *Air personality*

I taped a call-in and sent it to Pogo. I listened to the entire broadcast, or as much as I could after getting off the air. It sounded completely insane in the studio, and it was great to hear everybody calling in.

SHERMAN

Everybody started showing up; (there were) people I hadn't seen in a long time—Al Roker, Jr., Brooke Hunter—the funny thing was, I had never met Brooke Hunter. Brooke Hunter was there before I came in … here, we had never met, (but) I had known her name; I had listened to her when she was on Q101 a long time ago, and here she comes in. I'm not expecting her to say much to me. She comes in; she could not have been more sweet of a person. She's like, "Oh, I love your show. I listen to you guys all the time. I always wanted to meet you." I'm like, "Man, I listened to you." Brian "The Whipping Boy" showed up—he's one guy that I never thought would show up because he's with Eric and Kathy; I thought for sure that there's no way they're going to let him come over … Alex Quigley from (W)GN, he came over … I think Turd was on the line … James VanOsdol called; I was really shitfaced though by the time he called.

TIM VIRGIN

My only concern, and it may sound stupid to some people, (but) my only concern was making sure that we gave (Q101) the respect that it was supposed to get, and keeping 20 different personalities steered in the right way (not) to become negative. With Jim Jesus in the studio, that's not an easy thing to do.

JIM "JESUS" LYNAM, *Air Personality*

Whip came in on the final day of Q101. Me and Whip had this weird relationship when he was on the (Mancow) show; Whip did not like doing what he did. He wasn't a fan of being the Big Gay Butterfly, but he was *great* at being the Big Gay Butterfly. I got to tell him. He walked in and I was so happy to see him; I said, "Whip, I know you hated that character, but you were unbelievable at it."

DAVE BALL, *Air Personality*

Being in that room with all these people felt very special, like a unique club of people at the end of this great movie that people just watched, and we're all there for the standing ovation at the end.

ALEX QUIGLEY, *Air personality*

I was surprised at who was there, and who wasn't there. I mean, granted, the people—the current Q101 employees had already gotten the chance to say their goodbyes during the workday. And they went off to their own things. But the people who came back…I mean, I was sitting next to Sonic Boom Joey (Swanson), y'know? We'd never actually met!

It was Jaime Black and Chris Payne, the guys who no one else ever saw, but yet they were there week in and week out, and they were the ones who were chosen to be on at the very end. It was … it was kind of cool. I (also) noticed who wasn't there."

There were people who I thought would be there. And I'm not saying anything about the current -- like Sherman and Tingle or Electra, because they had already done their whole damn show. Alumni. Because, I know, unless I lived across the country, I would want to be there. I mean, I had to work the next day! I had to get a babysitter for Sam. I wanted to be there for that moment. Because we don't get to do that. That probably will never happen again. An actual, honest to God goodbye, that isn't watched over by management, finger on the button. That was…I remember me and Pogo, and Dana, jumping up and down, and, for some reason we started yelling, 'We won!' or 'We win!' —'We win! We win!' And I guess it was because we got to do it. We got to give the goodbye, the closure. And I felt like we did win. For once, the spirit of rock radio was real. For the last few days. I mean, I've seen the Arbitron numbers for those days. Granted, of course, they're going to be higher. It's like the Conan effect on NBC. But man, you couldn't deny that was also good radio. It was not bad, sloppy, college-level radio. It was professional, yet heartfelt and real.

SHERMAN

It was a very bittersweet end; it was almost like a high school reunion of people you didn't know, but kind of knew. It was something I had not expected, and I was so happy I was there for it.

FOOK

I thought it was some of the best radio that I ever heard. Even if you had never listened to Q101 before, even if you had only been a marginal listener to alternative, if you had grown up in that time, I think that was a valuable broadcast to listen to. It was a really, really good oral history of the radio station. Not to disparage (this book).

DANA LUCAS

Everybody was just standing around in the hallway, drinking, having a good time, hanging out, visiting with one another, taking pictures. It was just a good time.

CHRIS "PAYNE" MILLER

All the support staff was crying. Electra was crying. Tim Virgin was just funny and pissed off … there was a lot of hugging going on, but I think a lot of fear that I'm lucky to say I wasn't experiencing. It was the fear (of) people (who) knew that they were going to have to move from Chicago. A lot of people knew that they were going to have to sell their homes. A lot of people knew maybe they were going to lose their homes, or couldn't make their car payments. Everyone was going to lose his or her job. Everyone. It was very uncomfortable, but at the same time, although I felt for everyone else, I actually felt comfort in knowing that I was okay, that I was going to go back to my house. I felt pretty good about knowing that I was going to spend more time coaching soccer for my son, spending time with my wife, and taking my daughter to get her nails done, and just being a dad.

DAVE BALL

I understand the business side of it, why it happened. I don't agree with the decision to get rid of that radio station. Watching Chris Payne do his thing that last night, having JVO's stories go over the air, having Sludge call in from out of town, having Fook call in from out of town, having Alex Quigley being there, all of my marketing team, Jeff Delgado, all sitting there, kind of looking around, we all had a moment for sure. We were all shitfaced, just completely out of our minds drunk.

FOOK

For being a radio guy, it was a really comprehensive and entertaining broadcast about everything that happens within a radio station, and what a radio station means to a community.

But it was actually, probably, more for the people within the radio station, because it was kind of like a personal blow job for everyone there. It was super fucking cool to listen to.

CHRIS "PAYNE" MILLER

I decided I was going to play what were my favorite bands over the ten years that I did (*Local 101*). And that's what I did. I played some Smoking Popes; I played a Lucky Boys Confusion track; of course, I played Local H—"Hands on the Bible," the live song. That is just my favorite song ever—that version of that song, that album. That's just my favorite Local H track. And then I was told 'the morning of' that I could do two hours instead of one. That opened up another four or five bands that were on my list that I had stricken.

One thing I knew that it was important for me, once we concluded *Local 101*: I was going to play the Cure's "Friday I'm In Love." I had to. When we were in the studio, the computer program that we have in the studio, it was this computer screen you would see the name of the song and the name of the artist, on—what was it? Scott Systems or something. People were videotaping it and taking pictures of it, as (it) lined up, ready to go in queue, to be the last song that we were going to play. When we loaded it up, I didn't want it loaded up on the computer, until we were getting really, really close. As soon as we loaded it up on the computer … as soon as it went up there, people were saying, "Oh my God, thank you so much."

The next thing that I had to do was, I was going to let Jaime pick the very last *Local 101* song … Jaime was with me every show, and I'm not an easy guy to work with or work for, especially in radio.

Jaime was a great producer. He would set up interviews; he would prep the artists ... he never got any of the credit (and) he got all of the blame. He was with me from the very beginning, and I honestly do not believe the show would have been half as good had he not been there, and a part of it. It was important for me, in that last moment when I knew as many people (as possible) were going to be listening to it, that I acknowledged who he was, and that I let everyone know how important he was to me, that show, and that radio station, by letting him select the very last song on the show.

JAIME BLACK

Selecting the last song for *Local 101* in the final minutes of Q101 was one of those crazy, once in a lifetime-type opportunities that really only happen in movies. I never, in all of my time at Q101, ever imagined that that would be a bullet that I would get to shoot.

Chris suggested we close the final episode of *Local 101* with songs that meant something to us. I selected songs from Chicago bands that had made an impact on me. Acts like Spitalfield and The Fold. But even before I knew it would be the last song, I was dead set on "Tonight, Tonight" making it onto the final show. Large scale, the song was a love letter to Chicago by the largest alternative act ever to come out of this city. On a personal level, The Smashing Pumpkins were the band that first sparked my interest in local music, back with the release of Siamese Dream. Everything about it made sense, and in the hours and days that followed that broadcast, tons of people, both online and in person, applauded the selection. It had nothing to do with me; it was just the absolute right song to close *Local 101*. Billy Corgan even approved of the choice on Twitter the next morning. That was when the impact of that one song really actually hit me.

It's also worth noting that Chris has it all wrong here. It was not only an honor to work for and with such a stellar radio talent, but it was a lot of fun, too. Especially for 13 years! I absolutely owe my radio and music careers and experiences to Chris Payne, *Local 101*, and Q101, and the entire experience of being involved with the show is undoubtedly unlike anything I'll ever experience again.

POGO

I can forever be proud that I was there on the last day, in the last hours, in the last, really ... I got to introduce the final song with Chris Payne. It was all good, but that last day, that's the day you'll never forget.

CHRIS "PAYNE" MILLER

When that last song was over, literally, it was like the end of a rock concert. The lights came up really bright, everyone started filing out. There was this completely trashed control room. They went through about five or six cases of beer; I brought in three bottles of champagne. There was not much liquor. I honestly didn't see any drugs; it's this new generation. I honestly didn't see any. I don't think anyone was smoking cigarettes, either.

I had that … feeling of the end. That was it. That was the end. Then I walked out, grabbed my backpack, hugged a few people, and went and had some beers. That was it.

IN THE END: BONUS TRACKS

DANA LUCAS

I grew up listening to Q101. It was all I ever wanted to do, from the time I was 12 and the station first started. I knew I wanted to work for Q101. There was never a change from the very first day Q101 went on the air to the time I started working there. That was my sole goal in life: to work for Q101. I remember that feeling, walking out of the Merchandise Mart after I got the job, where I was like, "Oh, man, this is it. This is going to be amazing." And it was. Every day when I entered the studio, I still had that feeling.

JON REENS

If it hadn't been for Q101, my passion for music would have never been stoked, and my career path may have been completely different.

BROOKE HUNTER

First and foremost, my time at Q was most definitely the best time of my radio career. I was fortunate enough to be at Q at the height of the alternative music craze. I met some of my very best friends there, who I still keep in touch with to this day.

SHERMAN

Some radio jobs give you a career. I was lucky enough that Q101 gave me a family. I met my wife in my 10 years at the station, and had a son. All this was made possible by then-Program Director, Dave Richards, who gave me a chance back in 2001. From the bottom of my heart, thank you, Dave. Thank you, Q101.

DAVE RICHARDS

When I arrived in Chicago in 1994 as the Program Director of the brand new WRCX, my primary goal was beating Q101. Radio is competitive; the *"Art of War"* for market share. There's nothing I wouldn't do to win. I was coming to Chicago from Seattle, so I had a real good understanding of what "alternative" meant. At "The Rock," we were the bad boys of Chicago radio. But to be completely honest, Q101 in the mid-90s was the soundtrack of Chicago. There was simply something in the air with them. No matter what one-hit wonder they put on, it was a top seller in a couple of weeks. Credit.

ALAN SIMKOWSKI

Q101 was a very special place with great leadership, which is why it succeeded. We also made many friendships that will last a lifetime, and what could be more satisfying?

CHUCK DUCOTY

Managing a radio station in Chicago for a company that I continue to have a (great) deal of respect for, and being able to do it at a radio station that I thought was incredibly exciting, with a very high-profile morning, love him or not him … I love having high-profile talent. To me, that was just about a dream job in the business. I'm very, very blessed that I got to spend 4 ½ years at Q101.

KYLE GUDERIAN

To have the opportunity to work there, in that capacity, was the most fulfilling to me, and the reaching of a career goal, of sorts.

MADISON

The whole experience was exciting, because I came from … just shit radio … to the big time.

MICHELLE RUTKOWSKI

There was so much creativity and talent in that building. It was palpable. Some of the best in the industry have graced those studios, and I'm extremely proud to have worked with many of them.

POGO

There's a hell of a lot of people in my life now because of (Q101); there isn't a whole lot I would do over.

STEVE TINGLE

The people that I met out there on the street, who listened and supported the station; that was my highlight.

REY MENA

I've always viewed it, programming-wise, as schizophrenic. I think it was over-researched and lost touch with the people on the streets. It became more, musically, what was tested than what people really wanted. After so many vacillations back and forth, I think it lost the trust of the audience. Was it the Kid Rock station? Was it the Cure station? What was it? While the station was in the format for almost 20 years, if you compare it to stations like XRT, we talked about this when it hit the 10-year mark, we're saying, here's a station that has heritage. Does it really have heritage, or has it been in the market and the format for 10 years? To me, heritage is what XRT had. That whether it's 10 years, 20 years, what have you, everybody knew what XRT stood for. In retrospect, from a product point of view, it vacillated more than it should have. I think in doing so, it disconnected from the audience. At some point, the audience didn't even know what it stood for anymore. And I think that's what eventually brought it to its demise, more (than) the fact of not staying true to the essence of what alternative was. By that I mean, one day a band is really popular and they're the greatest thing in the world, and you've got to get behind it. Two months later, because somebody else starts playing it, they're crap; now they're on to the next thing. It's lost its connection. By comparison, an XRT built those relationships around an artist, and even though the artist may have fallen out of favor, they didn't fall out of favor with the artist. I think that wore thin with the audience. We became the flavor of the moment radio station; we did not support the artists. The artists we supported and loved and thought were the greatest things in the world, the moment somebody else played, we blew (them) off. And I think that was a big, big mistake.

JIM "JESUS" LYNAM

They should've stayed what they were, which was a cool alternative station. All of a sudden, we changed from being *the guys* to chasing the guys. That was never what Q101 was. It didn't matter who your morning show was. Who gives a crap?

ZOLTAR

I had some of the best experiences on and off the radio that will stay with me for the rest of my life. Those four years at Q101 gave me, like, mad confidence in myself, because I was questioning it when I got there; I was questioning it. It was cool. It was, like, it's hard to explain, but I did get spoiled. They took care of us. We were pampered. I can still remember those Christmas bonuses we would get. I would just be like, "God! More hookers!"

WENDY SNYDER

It was raw and real, that's what Q101 was.

SLUDGE

People in Chicago, and Midwest people in general, are, like, if you do something for me, I'm going to be loyal to you. That's what Q101 did all through the 90s and the early 2000s.

J LOVE

The greatest strength was—and I hate to sound cliché—the soundtrack of people's lives for that sustained amount of time, and it will forever be that to those individuals; and we're talking about millions of people over the course of 19 years.

MIKE STERN

It was an interesting place. There were just so many layers—it was so inconsistent for so long. Musical changes … from the Bill Gamble days of morning shows every, what, four months? to Mancow, and from Gamble to the opposite style of Dave Richards, to Tim. That sounds like it was a fascinating time to be there, just trying to steady the ship and put a hand on the rudder, and not sink us. I don't know that it was ever going to become a major powerhouse again; (there were) too many factions that remembered the good old days, too many factions that wanted to shake their fists and piss on the good old days. It was just … a unique place.

MIKE BRATTON

I met a lot of really good people, made a lot of good friends, and most importantly, I met my wife there.

JOEY "JUST JOEY" SWANSON

Every week was a highlight. I really, really enjoyed what I did there. And I'm truly proud of it as well. We played underground techno, house, and Drum & Bass on a station that was playing STP, Nirvana, and Sarah McLachlan … not to mention Metallica and Hanson. … and got away with it! I think that's pretty special.

TIM VIRGIN

I thoroughly believe, when you walk through the streets being a Q101 DJ that you realize how powerful that brand was, the first time around. I mean, holy fuck, man, this is Chicago. If you say, "I'm Tim Virgin from Q101," people know who you are. And in a big city, that's rare. When you're on the radio in New York City, nobody gives a shit about you because they don't love radio like the city of Chicago has grown to love radio.

It was an honor to work for that radio station. That's why, when I would see jocks coming in and out of there being, like, "Man, Q101 should be happy to have me," (I would think) "No, fuck you, man." You know?

EPILOGUE

As I was working to put this book to bed in May 2012, a sequence of events came together that I never could have predicted.

When Q101 signed off from its home at 101.1 FM in July 2011, Broadcast Barter Radio Networks, a small company run by two aggressive entrepreneurs (Matt DuBiel and Mike Noonan), purchased the station's intellectual property from parent company Emmis Communications.

That intellectual property sale included everything from the station's audio archives to the Q101.com domain. For months, DuBiel and Noonan worked to revive the Q101 brand as an online-only entity. From where I was sitting, it seemed like a problematic business model, simply because the station's image had taken such a sustained beating over the past nineteen years. I was skeptical about whether a corporate radio brand like Q101 could drive enough passion to sustain it as a digital enterprise.

I must admit, DuBiel and Noonan worked it hard, especially on the social media fronts of Twitter and Facebook. They even took to Kickstarter to try and fund a next-gen Jamboree. Using the crowdfunding site (which bankrolled this very book), they attempted to collect a whopping 299K, a total that included everything from talent fees to production costs. Ultimately, they got a respectable amount of pledges ($76,693), but that amount was more than 200,000 dollars short of the needed target.

As the online-based Q101's Kickstarter efforts were under way in early 2012, so too were the entirely unexpected machinations of Merlin Media, the company that took Q101 off the air in 2011.

In late April, news broke that Merlin had entered a local marketing agreement (LMA) with Venture Technologies Group LLC to take over the programming of smooth jazz-formatted WLFM-LP, a low-powered signal buried on the far left of the dial. Merlin replaced WLFM's call letters with a set of reactivated, familiar, calls: WKQX. The new format planned for the new WKQX? Alternative music. It was the most memorable "mulligan" of the modern Chicago radio era, especially when one considers that Merlin unapologetically put a bullet in the alternative format in 2011 and flash froze a handful of careers in unemployment lines. The new station, "Q87," was scheduled for launch on May 7, 2012.

This news clearly didn't sit well with the keepers of the Q101 brand, whose digital mortality was called into question in the face of Merlin's money and muscle. The day before Q87's scheduled launch, DuBiel put "Q101"-branded music programming on WJJG, the small suburban AM station where he serves as General Manager. In response, Merlin rushed to put Q87 on the air the same day.

Confused? Yeah, me too. It's pretty clear that the evolution of Q101 after July 2011, as well as the future of alternative radio in Chicago, is really somebody else's book to write.

CONTRIBUTORS' ESSAYS

The following essays were written and edited by people who pledged at one of two specific levels to this book's Kickstarter campaign. I'm thrilled to include their thoughts and memories in this section.

Because the authors paid to include their work here, I didn't edit or alter their writing in any way. Essays are arranged alphabetically by author within each pledge level.

Michael "Sheriff Scabs" Brya

Remember when Q101 was JUST a radio station? When it was just another preset between Z95 and The Blaze? Sure, it ended up being the soundtrack to my adolescence, but I believe Q101 had less to do with the music and everything to do with those who worked there. No other station ever made me halt everything to hear Way to Waste Time at Work, Quigley's Nintendo trivialities, Ryan Manno riffing on Juggalos, or Pogo's incessant affection for Pulp.

They provided enough memories to fill my own book, let alone this ten peso version. There's my frequent trading of *Princess Bride* quotes with Electra, Ginger Jordan calling to question my rumored dissolute encounter with Amy Lee (and Lynam asking, "How are her cans?"), my day in studio, when I discovered The Shuffle was a myth, Alan Cox abdicating promotional outings to attend my concerts, and lest we forget my bandmates drinking so much at the Morning Fix anniversary party, they swore off alcohol for a record six months.

But I'll always remember 7/14/2011, lying in bed, texting old high school chums, waiting to hear that last song. And the sullen feeling that followed, like a friend had died. But I'll also cherish the memory of that entire week, when Q101 became the very thing I always said it should be. A place where the on-air talent could display their extraordinary tastes, playing the music they loved, and sharing that love with us. It was the greatest week in Chicago radio history.

Ron Vanderhyden

For as long as I had been old enough to dial up my own radio station, my default and first choice had always been Q101. When circumstances kept me away, I would listen to something else, but always come right back without missing a beat. I had never thought about it. There was never a reason to.

Out of the blue one day I heard Electra's clearly emotional voice explain that Q101 would be going off the air soon. I discovered I was sad too. That drove me to think about things. In short, I found that what kept me coming back was the passion for the music, something I just did not feel from anyone else. I regretted taking this important gift for granted for so long. I listened as much as I could each day after that, as the station built up to its poetic end.

It takes strong people to celebrate during the loss of something so beloved. Doing so is to understand why you loved the thing in the first place, and to cling to what made it so special after it's gone. Music at its best is a shared experience amongst people that is so strong it can't simply be put into words. The final moments of Q101's broadcasting exemplified that power, and created a moment I was joyful to take in. I realized how thankful I should be for all the other moments I got to share in along the way.

Chuck Anderson

I started listening to Q101 around 1996. I was 11 years old. I'll never forget hearing bands like Nirvana, Live, and Everclear for the first time, not just because the songs were great, but because I wasn't really allowed to listen to that stuff then, which made them that much better. Discovering Q101 was like this weird treasure chest of music I had absolutely no familiarity with. It felt kind of taboo to, for example, stay up late to hear Korn's - 'ADIDAS' week after week at #1 on the top 9 at 9, but it made me feel cool in a way. The same kind of cool I felt spending so much time in junior high debating the other 98% of kids who liked B96. I hated those kids.

Craig Bechtel

Memories? Jim DeRogatis and Bill Wyman dropping my name on *Sound Opinions*. Winning Robert Chase's "My Three Songs" and getting free Blockbuster rentals for a year (answer: onomotapoeia). Before Mancow, there was Wendy and Bill. My ex-wife complaining about having to fetch coffee for "Join Me" Steve Fisher and how The Flaming Lips were weird—now she communicates with them regularly online. Q101 staff Christmas dinner, sitting by JVO. First "Twisted Christmas" concert, UIC Pavillion, including Veruca Salt (the background for their "Seether" video? Randolph St. Art Gallery), Bad Religion covering Christmas carols and a blistering set from Hole. Meeting Wyman's friend "Gina" there, only later learning she was Gina Arnold, who had just written *the book* on Nirvana. Farewell and rest in pieces, Q101.

Steven Boender

Loving, then hating, Q101 were the baby steps of any 90s adolescent music-snob in training. Elation that I could hear Sonic Youth, Smoking Popes and Morrissey on the radio was quickly followed by annoyance that other people could also hear those bands (MY bands!). Then Mancow, Fred Durst and "nü metal" provided a legitimate reason to hate Q101. Still, it was the gateway drug for so much amazing music. Packing my house in Oak Park in July 2011, shortly to leave Chicago for a new life in Portland (I know), I witnessed the death rattles of Q101, and felt a rush of a million awkward pimply memories. And it all ended with "Treason" by Naked Raygun. And it was perfect and sad and perfectly sad.

Joe Compton

It was Friday, and I was in love. Fresh out of high school and just discovering the world, the future was wide open; the soundtrack would be provided by Q101. I had always felt myself to be somewhat of an outcast, not quite fitting into any one category. What I would discover was that I wasn't alone. Whether it be rap, rock, punk or something totally different, Q101 played it all.

My taste in music was formed by what I would hear daily. But it wasn't just the music; the Djs made the station what it was. Irreverent and obscene, they rejected the mold of the typical DJ.

They made me laugh, made me think, and turned me on to music I had never heard before.

Melissa Dana

(Melissa didn't contribute an essay, though she paid to have one included in the book. As a sign of thanks, I've chosen to acknowledge her in this space—JVO)

Nick Hahn

My first boom box, walkman, car - so many mementoes of my life were audibly christened by Q101. It's sad that Q101 could have been saved by doing one simple thing: Play the music that people love most.

You see, every holiday season Q101 played the top 101 songs from each year it existed, then letting the listeners vote on their favorite year. Almost always 1994 won. No surprise with the top songs being from NIN, Soundgarden, Offspring, Live, STP, Weezer, Green Day, Beck, Pumpkins...and the list just keeps going.

So what did Q101 do about this resounding outcry to play more of these songs? They played obscene amounts of Limp Bizkit, Nickleback, and other flavor-of-the-moment-but-not-so-great music. :(Even so, you're greatly missed.

Ben Husmann

I worked in the Q101 office (not for Q101 directly) from 2001-2005. Because of my proximity to the wonderful world of radio I was able to have the following unique and amazing experiences: Cinco de Mayo margaritas and tacos at 8AM, photo ops with Mr. T, shaking hands with a chimpanzee, holiday breakfast complete with jello shots and made-to-order omelets, Friday morning drunk club girls clogging up the halls, free tickets to countless shows at the Metro & House of Blues, office bathrooms covered in feces, endless free lunches, skyboxes at Sox games, and hundreds of dollars of game play at Dave and Busters. Now I have to pay for this stuff but the bathroom walls are clean.

Curt Mazur

I never really had a great taste in music as a child, I just didn't like any genre, no beat pulled my heartstrings. In 05' my dad passed and I got to play with the garage radio, one day I came across *Santa Monica* on Q101 and loved it. I found myself in 90's alternative. The last day Q101 was on air I had to leave town, didn't want to but had to. Fifty miles down I-55 every weekend, I knew I'd lose Q101 by the time I'd hit I-355 and I died inside. But for once in five years the sky was perfectly clear as I listened to Electra all the way out to Braidwood, IL. God let me listen to heaven that day.

Carolyn Notorangelo

Cut the speakers, when you see his leaf fall, when aid and amends weep, fraught for melodic destination.

Cut the speakers, at the soft landing, as reverberating tones ring supernatural and untouchable.

Temporal security is distraught. The musician is dead.

Call out if you will, out for memory of a ludicrous and falsified past. His pure notes will cut fattened minds. His melodies trace the follicle-filled bumps that sprout on soured spines. Late at night, the plump assets of gain will reveal themselves incomplete…not to musicians, but to artists of deceit.

Beyond distant empty studio walls, a single surpassed soul will echo the absence in set minds.

The silence; the death of a musician.

Andrew O'Connell

"I fear that I am ordinary, just like everyone." This line from Muzzle by the Smashing Pumpkins, a song Q101 frequently played in 1996, embodies the attitude of Q101 and why it was important. My Q101 best/worst memory was trying to be caller 101 to win a spot in a ZWAN video, but when I was indeed the correct caller, I learned I had won Linkin Park tickets instead. I was one disappointed Corgan-aholic.

Also, I had my 13th birthday party at the Adler planetarium to watch music videos on the ceiling after winning tickets from Lance & Stoley, the morning DJ's. Strangely enough, my parents still have a photo of Lance, Stoley, and my 6 year old brother in their living room (weird).

Jon Reens

Working at Q101 during its heyday forever changed my career path, by showing me that you can truly love what you do for a living. My time in promotions paved the way for my positions at Ala Carte Entertainment, House of Blues, and ultimately to where I am at Live Nation. Who would have thought that working liquor promotions until the wee hours of the morning, or working on events such as Jamboree and Twisted Christmas, would have led to me finding my career. RIP Q101.

Colin William Reed (2004)

Dear Q101,

Call me idealistic, but I believe people like to listen to good music, meaning not created by artists "constructed" by record companies simply to make them ungodly amounts money. I understand it's the music *business*, but it's *also* possible to succeed without damaging the credibility of the medium.

That's the reason why I desperately want to work at Q101: you're not a bunch of soulless sellouts. As a Chicago native, I know you've strengthened the local scene by playing great material, sponsoring quality live shows and giving hometown bands airplay.

So, why am I any more deserving of this internship? I promise to relentlessly work my poor student butt off. That being said, I look forward to being your coffee bitch. Thank you.

Apryl Sessoms

Chicago's Q101 positively impacted my life defining who I am today; its music was in the background of my fondest memories growing up. At thirteen, I'd sit on the floor with my radio, a blank cassette, and a finger ready on the "Record" button. That was 1995, and how I started my Q101 music library. The following summer, my dad bought me my first three CDs: Evil Empire, Sixteen Stone, and Tragic Kingdom. As young teens, my sister and I would rollerblade in the garage, never straying too far from Q101 over the radio. Over a decade and a half later, Q101 continued to play my favorite songs, introducing me to new favorite bands along the way. Music is my LIFE; Q101 played the soundtrack.

Chris Stiles

I was hired to be part of the interactive crew that worked out of the back of the Q101 studios. I was straight out of college, I had just moved to Chicago from Florida, and I lucked into this job at what seemed to be a reasonably cool radio station. For the first time in my life I was going to earn a real salary, with real benefits, in a professional work environment. I walked through the employee entrance on my first day and was stopped in my tracks by a wall of pot smoke. Cypress Hill was sitting in the Green Room twenty feet down the hall, smoking up, and I thought, "Wow. This is it. This is how my first real job starts."

Lance and Samantha Tawzer

We met because of Q101. We were engaged because of Q101. We have our proposal on a CD because of Q101. We have two gorgeous sons – Evan and Henry because of Q101. We have a lifetime supply of key chains and T-shirts because of Q101. We were given advice on how to have a successful marriage by Johnny Rzeznik of the Goo Goo Dolls because of Q101. (He later divorced – we didn't.) We have many brilliant cocktail party stories because of Q101. Once in a blue moon someone will say "…didn't you used to be…" because of Q101. Mostly we had some phenomenal experiences, made some life-long friends, fell in love and simply are the little family we are because of Q101.

James Weber

There are very few words that can describe how I think about Q101. In short, Q101 meant I was home. During my formative years, I was living outside the range of the station and whenever I was on my way back home, I knew I was getting close I was able to receive Q101. I knew I was close to my family, friends, past good times and future good times ahead. The loss of Q101 is not limited to the loss of a great music station, but an outlet for people to discover new bands and new types of music. The void created by the loss of Q101 will never be filled and the memories created will never be lost.

Stacy Zielinksi

    I truly believe my life started in 1995 when I discovered Q101. I can't help but think how much I owe the station for playing bands like No Doubt and 311 on rotation and how dramatically my life shifted from that year forward. High school brought the days of Zoltar's Industrial Zone and Chris Payne's Local 101, imagining those years without it seems impossible. Mancow always made me excited to start my day. Years later, Sherman and Tingle picked up where he left off. Sludge's Road Rage may be the only time my evening commute was bearable. Jamboree's, Twisted's, Block Party's created endless memories I'll never forget. My life is now music and music is life, and no one is to blame more than Q101.

THANK YOU

This book is the result of the generosity of hundreds—literally hundreds--of people.

In June 2011, I launched a 'Kickstarter' campaign to write and publish this book. The funds generated on the crowd-funding site, as I explained in my pitch, would support everything I needed to make We Appreciate Your Enthusiasm a reality, from copy-editing to design and formatting costs. One month into the effort, I found myself successfully funded, humbled, and grateful.

The contributors below each pledged money to this book in exchange for a reward that included a mention on this page. I'm thankful for their contributions, and hope you'll take a second to acknowledge their efforts by reading through the following list.

Dayna Abel, Emma Abraham, Barry Allen, Jake Alleruzzo, Mike Allison, Amanda M. Andersen, Crystal Arevalo, Matt Arnold, Marie Bajza, Casey Baker, Andrew Bannon, Patrick Barnum, Angelique Barthel, Guy Bauer, Brenda Bernotus, Natalie Beyer, Jeremy W. Bick, Eric Binder, Melissa Birch, Kevin Birtcher, Ryan "Elwood" Bjorn, Blue Damen Pictures, Alex Borring, Mandy Borvan, Matthew Boudnek, Kyle A. Boyd, Steve Boyd, Matthew Brands, Ryan Braschko, Mike Bratton, Dave Broche, Matthew A. Brooks, Patrick Brower, Sean Cable, David Calabrese, Kevin Callero, Mike Caraker, Frank Carrillo, Brian Carmack, John Peter Caruso, Kevin Casper, Niles Caulder, Cory Chism, Jeff Christoff, Ben Chung, Cinchel, Hannah Jo Clark, Jeff Clayton, Jeffrey Cobia, Clint Cochran, Katie Colaianni, Greg Corner, Katie Crain, Gabe Culberg, Letitia Cullens, Ryan Cunningham, Arthur Curry, Sidarth Dasari, Neil Dattani, Chris Deckinga, Nick Demetralis, Adam Dickens, Pete Dillenburg, Wayne Dixon, Paul Doraski, Timothy Drake, Adam Dremak, Cass Drew, Anthony & Fran Dubicki, Chris Dumsick, Daniel Dumsick, Cheryl Durlak, Jan Engle, Olivia Espinoza, John Farneda, Eve B. Feinberg, Andy Finnigan, Ryan Fisher, Walter Flakus, John Flickinger, Justin

Follis, Irving Forbush, Nikki Foy, Gabriella Frate, Michael Freedman, Brad Fuhr, Gwydhar Gebien, Jocelyn Geboy, AnnMarie Genovese, Harrison George, John Giannerini, Amanda Giarratano, J.P. Gleason, David J Glick, Jessica Goforth, Marty Golenzer, Eric Goldesberry, Eric Gomoll, Rob Goodman, Justin Gorrell, Paul Grachan, Nick Graf, Mike Grant, Michael Grigsby, Benjamin J. Grimm, Walt Grogan, Phil Grosch, Joe Gunia, Timothy B. Hailey, Carter Hall, Crystal Halm, Scott Hames, Ryan Hanson, Roy Harper, Chris Hauger, Shawn Headrick, Derek Hebal, Keith Heffernan, Keith Hellmann, Aaron Hellwig, Diane Heuring, Ryan Hilsabeck, Kevin Hilty, C.A. Honore, Ammie Hoppenrath, Colleen Horcher, Natalie Horcher, Rachel Hostert, Noel and Becky Janssen, John Jenzeh, Jeffrey J. Jett II, Kevin Johanson, Sean & Michelle Johnson, Sarah C Johnson, Tim Johnson (former Q101 Marketing Director), Tracey Johnson, Brian Jones, Lauren Daye Jones, Danielle Jozwiak, Felix Jung, Elizabeth Jurcik, Tony Kaczanowski, Jennifer Kazin, Phil Kapocius, Aaron Keefner, Adam Keene, Patrick "Spanky" Ketza, Jeff Kitsmiller, Jr., Ron Klawinski, Stefanie Kljucaric, Christopher Kois, Christopher Kolar, Rachel P. Kreiter, Mary Kroeck, Froggie Krom, Christine Kwon, Melanie LaForce, Matthew J. Laird, Adam Lamey, Anthony Lampl, Tom Leddy, Paulina Leduchowska, Brian Leli, Kelly Lence, Jeffrey Lesniak, Steve Levy, Brian S. Lewis, Kyle Lewis, Laura Liekis, Andrius Lietuvninkas, Andrew Linder, David Loechle, Jen "Jameson" Longawa, Nick Loxas, Joel Jordan Lozada, Jennifer Ludwick, Patrick McCarron, Adam McCrimmon, Erin McGowan, Justin McMillen, Szymon Madej, Sandra Madison, Mike Malinowski, Shawn Mangan, Riley Mangan, Ryan Manno, Kevin Marinier, Patty Martin, Anna Mattson, Christopher Mazella, Maria Medina, Justin Melcarek, Alyson Merkner, Tim Messer, Brucha Meyers, Dan Milano, Mark Miller, Stephanie Modica, Mike Morkin, Steve Moudry, MrJM, John Mucha, Matt Murdock, Nicole Munoz, Maria Nedina, Thomas Negovan, Chris Nelson, James Nelson II, Kent Nelson, Tom Niedringhaus, Tracy Nemitz, Jeffrey Niebres, Elizabeth Nolan, Brianna North, Barry O'Connor, Kyle O'Laughlin, Christine O'Rourke, Jason Obaob, Joey Odorisio, Javier Ontiveros, Ray Ortiz, Paul Ostaszewski, Nicholas Otte, Pamela Pahnke, Ray Palmer, Jackie Pancotto, Angela Papaleo, Hailey Parejko, Frank "Paco" Partida, Dana Parz, Todd Michael-Bentley Paulson, Amanda Peterlin, Laura Petraski, Bradley Petrik, Jenny Pfafflin, Shane Phillips, Bradley Piasecki, Kathryn Powell, Andrew Powers, Kitty Pryde, Mary Jo Quinones, Ken Proctor, Jason Radke, Daniel Ramirez, Juan A. Rangel, Ryan Rau, Joseph Redeker, Dan Rehberg, Roberto Reyes, Brian Rhodes, Reed Richards, Jim Rick, Christiane Rodes, Joseph Romanowski, Margaret Rooney, Adam Roppolo,

Amy Ross, Paul Rossi, AJ Rudolph, Kimmi Rudolph, Michelle Rutkowski, Jack Ryder, Mike Sakiewicz, Brian Salata, Frank Salemme, John Sales, Jose Sanchez, Melody Santos, Roger Schlogel, Carrie & Eric Schmidt, Russell Schoenbeck, Alex Schwarz, Sean Patrick Sengenberger, April Sessoms, Chris Sevening, Julia Shell, Brian and Katie Sherman, Mike Sherry, Justin Siddons, Timothy James Sipple, Brad Sliwa, Chad R. Small, Kevin "trkdrvr" Smith, Laurence Smith, Erik "SWIRLY" Sorlie, Bob Sparks, Spencer G. Spathis, Jenni Sperandeo, Jack Stachowiak, Cliff Steele, Hannah Steele, Kara Steiner, Anthony Stephens, Mike Stergos, Josh Stoyak, Mike Surerus, Claire Suriano, Katie N. Suvada, Paul Swanson, Stephen Swintek, Stephanie Tappan, Ben "Dreads" Taylor, Andy Tihonow, Staci Tipsword, Skip Tramontana, Jessica Tuchowski, Ken Turke, Nick Ulczak, Tony "Bags" Valdez, Eddie N. Villarreal, Christie Wagner, Christian Wagner, Luke Walker, Brian Ward, Josh Watson, Kristina Weld, Wallace West, Kathy Whisler, Paul Wilk, Jeannine Wisniewski, Rob Wong, Melissa Wood, Dave Wydra, Ben Young, Katie Zabielski, Tony Zaino, Zatanna Zatara, Mike Ziberna, Matt Zivich … and that son of a bitch, Tony Bosco.

ACKNOWLEDGMENTS

Thank you to the hundreds of employees, interns, and miscellaneous personnel who made Q101 a successful FM station in the ultra-competitive Chicago radio market for close to 20 years. Your contributions gave me a story worth telling.

Though it would have been impossible to include everybody who's ever punched the clock at Q101 in this book, their presence in the stories and memories you just read is implicit and appreciated.

To properly tell the oral history of Q101 between 1992-2011, I spoke with over 75 intelligent, prolific, and funny raconteurs over a nine-month period, beginning in July 2011. I'm forever grateful for their time and candor. On a much more selfish level, I'm thrilled to have had the opportunity to reconnect with so many old friends and make connections with new ones.

I'd like to thank a group of friends and allies whose help and wisdom helped make this book better than I could have imagined it: Mike Bratton, Mike Englebrecht, Patrick Brower, Marty Golenzer, and Jaime DeMedici.

Thanks to Tim McIlrath for contributing the Foreword. I'm a Rise Against fan, and feel truly honored to include his work in my book.

Thanks to the talented Meagan Evanoff for her work in formatting the manuscript book's interior and making it look awesome and print-ready.

Special thanks to Kyle Baker (bakerprints.com) for designing *We Appreciate Your Enthusiasm*'s amazing, suitable for framing, book jacket. Watch out for the burger!

Thanks to Steve Dahl, for doing everything first.

Work on this book threw me into a fairly demanding routine. Every night, after putting my kids, Noah and Fiona, to bed, I'd camp out in front of my computer to slowly, methodically, work on the book's interviews, manuscript, and fine-tuning. My wife, Anne, deserves great appreciation for understanding my dedication to the project, and for unconditionally supporting me throughout.

Noah and Fiona were also very understanding when I had to disappear to work on the book on random weekend afternoons. I look forward to the day when I can let them read this book. Maybe after high school.

Made in the USA
Columbia, SC
16 June 2025

59456500R00193